## Dedication

I am dedicating this book to my wife, Bina, who has stood by me for more than 50 years and is the rock of the family. I also dedicate it to our family—our two lovely, accomplished daughters, Sonia (a clinical psychologist) and Priya (a banking professional); our two sons-in-law, Steve (a scientist and intellectual property agent) and Manmeet (a mental health professional); and our four grandchildren, Diviya, Simrin, Preeti and Arjun, who will surely enjoy the benefits of AVs much more than the rest of us. Diviya's has ambitions to become a strong leader and ensure women are equal to men in every way. Simrin is a budding robotics and AI scientist. Preeti will surely come up with some new AI-based method of non-invasive neurosurgery when she grows up. And Arjun—the youngest in the family—will follow his mother, a banker, and take the Canadian financial system to higher levels of internationally recognized stability. I see in them change for the better, hope for prosperity and progress for humanity. They energize and encourage me every day to keep contributing to society in small ways. This book is for all of them.

Suite 300 - 990 Fort St
Victoria, BC, V8V 3K2
Canada

www.friesenpress.com

**Copyright © 2018 by Chander Dhawan**
First Edition — 2018

The author acknowledges, without reservation, the writers, business journalists, academics, researchers, consultants and professionals who have published their insight on the topic. In many cases the author has quoted from the sources directly. The majority of the images come from vendor websites. The author thanks all his sources.

The author has made every effort to acknowledge all of his sources and seek explicit permission wherever it was possible. He apologizes for any omissions, which are not intentional. Please bring any issues to his attention at chander.dhawan@rogers.com. He will correct them in a subsequent edition of the book.

All rights reserved.

No part of this publication may be reproduced in any form, or by any means, electronic or mechanical, including photocopying, recording, or any information browsing, storage, or retrieval system, without permission in writing from FriesenPress.

ISBN
978-1-5255-3982-4 (Hardcover)
978-1-5255-3983-1 (Paperback)
978-1-5255-3984-8 (eBook)

1. TECHNOLOGY & ENGINEERING, AUTOMOTIVE

Distributed to the trade by The Ingram Book Company

# PREFACE

The idea to write a book on autonomous vehicles (AVs) came to me about two years ago. I've always tried to stay up to date with what was happening in the smart phone and telematics industries, receiving regular newsletters and keeping tabs on news stories. As time went on, I, along with everyone else, realized that "self-driving" had become a buzzword in the media—there was extensive coverage, not just in the automotive press but everywhere. There were a few books written about the topic as well.

However, I realized that the majority of the content out there was simply echoing the PR messaging from the various industry proponents. The conversation was not very comprehensive or holistic either – certainly not in one place. It didn't adequately cover all the perspectives, and it lacked critical analysis. I knew there was much deeper knowledge and expertise out there, but it didn't exist in a single book that was easy to understand and that provided sufficient technical depth without getting consumed by the intricacies of sensor technology or elaborate AI algorithms. I saw a need for a sort of primer for technology analysts or market-development professionals who wanted to understand the broader picture—from 32,000 feet. After reading such a primer, they could go and dig more deeply into any area that interested them.

That's what I've tried to create with this book.

What qualifies me to write a book on autonomous vehicles? First, let's consider some of the qualifications I *don't* have. I'm not an automotive or autonomous vehicle engineer who can explain the intricacies of automobile components or of how to turn an ordinary vehicle into an AV. I'm not an academic researcher from MIT, Stanford or CMU who can explain the latest developments in artificial intelligence. I'm not an automotive business strategist or auto industry analyst, either. I'm not a venture capitalist specializing in the AV sector with insight into where the smart money is going in search of AV opportunities. I'm also not an automotive astrologer who can accurately predict when some of the AV developments will actually happen.

So, if I'm not any of these qualified AV experts, what qualifies me to be here at all? What I bring is the background to understand and explain the big picture for AV development and adoption. I'm an engineer, a management scientist and an experienced technology practitioner—with a pretty decent understanding of technology-development and business-development processes. I started my computer science career at IBM in 1969. My first job was to transfer leading-edge technology solutions from the lab to the marketplace. Throughout my career, at IBM and later as an independent consultant over a span of 40 years, I implemented many "state of the art" IT projects, including the largest and fastest (at the time) real-time financial network in the world.

I developed a deep understanding of the intricacies involved in taking a technology from the lab out to the public. In the early 1990s, I grew fascinated by wireless technology and mobile computing, and I became a recognized industry expert. I wrote two books on systems integration for mobile computing technology and wireless internet. Practitioners and academics read my books to learn about building end-to-end solutions using the component technologies they developed.

I also spent a lot of time advising public-sector clients on how to implement real-time driver and vehicle systems as well as police systems for dealing with the driving public. I also advised the Ontario Government on public insurance issues. I have an excellent understanding of the system-automation issues that the custodians of our state and provincial road infrastructure face.

My career and experiences have brought me right to the intersection of engineering, IT, emerging technology and public-sector driver-related issues—exactly where we find the AV industry. And *that's* what qualifies me to take a broad, system-wide look at the world of autonomous vehicles.

### An Open Acknowledgement for the Content

I spent two years researching this book, and I reviewed over a thousand articles, reports, vendor websites and books. I want to take a moment to acknowledge, without reservation, hundreds of writers, business journalists, academics, researchers, consultants and professionals who have published their insight on the topic. Their writing has helped me in creating this book, and in many cases I've quoted from the sources directly. The majority of the images I've used come from vendor websites. I owe these authors and vendors gratitude for the knowledge that they have given me and for the value they have brought to my book.

I have made every effort to acknowledge all of my sources and seek explicit permission wherever it was possible. I apologize for any omissions, which are most definitely not intentional and for which I take full responsibility. Please bring any issues to my attention at chander.dhawan@rogers.com; I'll be happy to correct them in a subsequent edition of the book.

## My Personal Bias
I'll also disclose my personal bias. I'm an engineer-turned-management-scientist and mobile technology consultant who believes in the important role of emerging technologies in our lives. I believe that humankind must continue to evolve and adopt new ways of doing things. There are many factors that bring change to our lives; technology is one of them. I believe we should accept change if it makes our society better.

I'm also a 70 plus baby boomer. I've seen a lot of change over the course of my life. I believe that the pace of change should be managed so that we can adapt to new ways of living at a natural pace that doesn't cause anxiety and disruption to large segments of the consuming public. My professional life has involved implementing emerging technologies rather than creating them, so I'm also a realist, not a dreamer. Over my long career, I've come to realize that if technology is forced on people, they resist forcefully. I've learned to be humble and allow change to proceed naturally.

## My Book—A Comprehensive and Critical Analysis from a Holistic Viewpoint
This book reflects opinions I have come to form, after reviewing the huge amount of available information and talking to a variety of industry professionals, on the state of the AV industry today and its outlook for the future. I've tried to present a balanced, holistic viewpoint of this new technology, which has the potential to significantly disrupt the auto industry. I've described some of the fundamental principles that are more likely to govern the adoption of this technology than the industry's marketing drive. I've even recommended a methodology for arriving at balanced decisions about the industry, and I've suggested that regulators and legislators try this approach and debate whether the industry's pace of change is in sync with society's desires overall. My goal is to provide independent advice to all of the industry stakeholders, who together will determine the pace at AVs take over our driving lives.

## Repetition is Intentional
Many of the topics in the AV world are inter-related, and are intentionally covered in multiple places in the book to make sure each section can stand on its own, knowing that some readers may find some sections more relevant to their interests than others.

## Professional Acknowledgements
On a professional level, I must thank my editor, Ms. Jess Shulman, who has cleaned up my original manuscript so much and so nicely that sometimes it looks more like

her work than mine—in a good way! Not only did she correct my grammar but she also rewrote awkward sentences and paragraphs to make them much more clear. She also gave me lots of great advice about how to properly cite my sources. Thank you, Jess!

**Proceeds from the Book are for Charitable Purposes**

I've enjoyed a successful career and now I want to give something back. All proceeds from this book, after recovering publication costs, will be donated to charitable causes of my grandchildren's choosing (whether it's to pay for a child's education in Africa, to encourage young Canadian aboriginal girls to study STEM [Science Technology Engineering and Mathematics] subjects or something else, they will decide). Autonomous vehicles are for our future generations. As members of Generation Z, my grandchildren will have a whole different way of (driving) life than I've had, and I am very grateful to have played a small part in describing that future through writing this book.

# Recognizing Key Innovators of the AV Evolution

I am recognizing the following five individuals, one academic institution, two research organizations, one technology company and one auto OEM who have moved the AV industry forward. This selection is based on my personal research and is purely subjective.

**Prof. Ernst Dieter Dickmanns** is a German pioneer of dynamic computer vision and of driverless cars. Dickmanns has been a professor at Bundeswehr University Munich (1975–2001), and visiting professor to Caltech and to MIT, teaching courses on "dynamic vision". He was involved in pioneering research in Europe that led to the eventual development of ADAS and semi-autonomous features in European cars especially Mercedes. **Photo Credit: www.wet.de**

**Prof. Hinton** is a Canadian researcher of British origin at the University of Toronto and working at the Google Brain lab in Toronto. He is a leading figure in the deep learning community and is called by some as the "Godfather of Deep Learning" whose research truly enabled application of computer vision and AI to mimic the human driver by software-based autopilot. He continues to refine this unfinished effort. **Photo credit – University of Toronto Press**

### Sebastian Thrun – Project Leader at Google for Self-Driving Car Program

The self-driving project was the brainchild of Sebastian Thrun, the 51-year-old former director of the Stanford Artificial Intelligence Laboratory, a Google engineer and the co-inventor of the Street View mapping service, Sebastian converted the known research and Google's research platform into a self-driving car trial that took the auto innovation leadership away from the established hundred year old OEM community. He is currently the CEO of the Kitty Hawk Corporation, chairman and co-founder of Udacity. **Photo credit – Google**

### Elon Musk – an Innovator

We recognize Elon Musk for being a visionary and a doer in the electric car and AV industry. He founded Tesla and was personally responsible for leading the architecture, design and building of the first electric car with semi-autonomous features and a range of 300 miles. He implemented key features of AVs in Tesla model S. Irrespective of all the problems of manufacturing Tesla model 3 cars in quantity, broken delivery promises, achieving profitability and investor/SEC controversies, Elon will go down as a pioneer in electric AV saga. **Photo Credit – Tesla**

**Amnon Shashua - Mobileye CTO:** Amnon Shashua is a computer science professor at the Hebrew University in Jerusalem as well as co-founder and CTO of Mobileye. Under his technical leadership, Mobileye converted a lot of research from Europe and USA into practical ADAS solutions that were installed in millions of cars around the world. **Photo Credit – Intel Mobileye**

**US Govt. DARPA Office:** It was only after US Government's DARPA Challenge that US academic institutions and Google itself became serious about implementing results of research as of 2005 into real-life AV prototypes to understand what was possible and what more had to be done to realize the goal of self-driving AVs. Interestingly, some of the winners of this challenge companies like Google and started industrial projects.

**European Eureka Prometheus Project:** Much before US took over leadership of research and implementation effort for AVs, Europeans mounted a massive 850 million Euro research project across multiple nations. Prof. Dickmanns of Germany and Prof Brogio of Italy were key figures in this effort. The project achieved excellent results and executed outstanding trials on European autobahns.

**CMU:** While MIT and Stanford University are world class institutions and have done extensive research in AV related field, we believe Carnegie Mellon University in Pittsburgh should be given the distinction of spearheading the application of known technological knowledge and experience into prototypes or modified vehicles into autonomous vehicles. Some call CMU as the birthplace of AVs as we know these cars today. In fact, CMU efforts during 1980s represented second stage of the development of AVs wherein multi-disciplinary teams (engineers - mechanical and electronics, robotics experts, AI neural network researchers and systems integration practitioners) cooperated to create a product that could be put to real-life trials.

### Google/Waymo

We would like to give Google/Waymo the credit for capturing an opportunity that showed up after DARPA Challenges. Google has demonstrated a consistent effort towards making self-driving car a reality after 2009. Their deep knowledge of building computer hardware/software solutions based on existing and non-existent technologies (like AI) for complex engineering problems. Their AI-based deep learning knowhow, establishment of a simulation lab and research into reducing the cost of sensors will go a long way in making affordable AVs of tomorrow.

## Daimler

Out of all the established OEMs, Daimler Corporation of Germany (manufacturer of Mercedes brand of cars) can be given the unique distinction of participating in early trials of autonomous cars. Daimler was deeply involved in Prometheus project of Europe. Even today, it has been conducting trials in both ADAS-supported implementations since 2015. It is also involved in higher levels of SAE level 4 trials with an eye on level 5. Daimler has an advantage in producing expensive cars where cost of sensors and software upgrade is a lesser issue in other brands.

# TABLE OF CONTENTS

Chapter 1 — "AV Plus"—Setting the Stage..................................................13

Chapter 2 — Fundamentals of Technology Innovation and Adoption................33

Chapter 3 — AV History—How We Got Here...................................................43

Chapter 4 — The Rationale for AV Plus...........................................................65

Chapter 5 — AV Plus Functional Component Design........................................77

Chapter 6 — The AV Plus OS, the Human-Machine Interface and AI..................93

Chapter 7 — The AV Plus Ecosystem .............................................................115

Chapter 8 — TaaS—Concept and Players......................................................143

Chapter 9 — AV Perspectives, Part 1 (Consumers, Suppliers, Academics).......157

Chapter 10 — AV Perspectives—Part 2 (Insurance, Legal, Regulatory)............175

Chapter 11 — A Methodology for Balancing Multiple AV Perspectives.............187

Chapter 12 — Challenges Facing the AV Plus Industry...................................195

Chapter 13 — Opportunities for Entrepreneurs and Startups..........................217

Chapter 14 — Crystal-Ball Gazing - What, When, How - Evolutionary Path.......225

Chapter 15 — The Way Forward for AV Plus..................................................247

The Last Word – No, an Interim Word............................................................263

The Index ...................................................................................................265

# Chapter 1

# "AV Plus"—Setting the Stage

*Autonomous vehicles (AVs) are everywhere in the media, but they're not in dealer show rooms yet. A few prototype AVs are roaming around on public roads in Phoenix, San Francisco, Pittsburgh, Toronto and elsewhere. Consumers are excited but also a little puzzled. They're asking: when and how? Much progress has been made technologically to produce early versions of fully autonomous cars. Sophisticated sensors are getting more and more refined, becoming cheaper and smaller. "Deep learning" artificial intelligence algorithms are giving these vehicles the brains to assist (today) and replace (tomorrow) the human driver. Huge amounts of VC money and investment are being put on the table. Avant-garde vendors from Silicon Valley are giving the stalwarts of the industry a run for their net worth.*

*On the surface, it all seems very exciting and promising. We could see all kinds of societal benefits—traffic congestion alleviated, even lives saved. Auto companies have already announced aggressive launch dates for AV ride-hailing services. When and how consumers' (especially millennials and Generation Z) dreams will become a reality remains an open question. The potential rewards are significant, but stakes are high and challenges are many. The industry will see major changes, but it may take longer than the promoters would have us believe.*

## 1.1 Some AV Trivia

Let's begin with some trivia about autonomous vehicles and the artificial intelligence that powers them.

1. Over 1.2 million people die every year in car accidents—more than in wars around the world.
2. The first attempt at a driverless car was made in 1925 in New York City.
3. The first cars, before the internal combustion engine was invented, were electric.
4. The average private car is used only 4% of the time—the rest of the time it sits idle in the garage or a parking lot.
5. If auto makers wanted to build reliability into future cars and cost was not a consideration, these cars could be five times more reliable than today's cars. (Commercial airplanes are used, on average, eight to ten hours a day and have a lifespan of 30 to 40 years, unlike today's cars, which are used for two to three hours a day and last only 10-15 years.)
6. If we used AVs as public transportation, vehicle ownership could decrease significantly, as long as the cost was lower than private ownership and the service was convenient. Some cities could ban car ownership altogether, just as Singapore is planning to do now.

7. Transportation contributes 50% of the world's carbon pollution. With widespread use of electric cars, carbon output could be reduced drastically.
8. According to Bloomberg, 54% of all new-car sales and 33% of worldwide fleet vehicles will be electric by 2040. Ultra-fast charging technologies that will "fill 'er up" in 10–15 minutes are on the horizon, and should be available by the time AVs become mainstream.
9. Sensors and cameras capture distance information about obstacles and images of the road in terms of digital pixels. The current state of the art in sensor fusion of this information and deep-learning AI techniques still can't give us the detailed information about roads, surroundings and obstructions that we get from our eyes. There will need to be more refinements in sensors, HD mapping and AI in the years to come before a robotic autopilot can see as clearly as a human driver.
10. AI is not a new branch of computing; it's been around since the 1950s. But it's only now that computers fast enough to handle AI's processing requirements are coming into prominence.
11. AI for AVs is the most complex application computer scientists have tackled so far. Some experts, such as Lance Eliot of Cybernetics and Tim Cook of Apple, equate it in complexity with the moon-shot efforts of the 1960s.
12. AI talent is scarce. AI developers are being paid $300,000 a year if they have a Ph.D. with no experience or a master's degree with five years' AI experience.
13. China is investing heavily in AI. The country could overtake the United States in the future.
14. Europe was ahead of the United States in AV research until 2000, when the US government pumped funds into Carnegie Mellon University, MIT and Stanford. The DARPA challenge in 2005 woke the Americans up, and Google's self-driving car project and Tesla's Autopilot have crystallized the commercial effort.
15. Google's autonomous cars have driven eight million actual miles and several billion miles in simulation, yet the prototype autonomous cars still don't understand the driving world nearly as well as humans do.
16. Power and computer real-estate requirements currently leave no trunk space in AVs. A lot of miniaturization is required—something Silicon Valley is working on, of course.
17. AVs accumulate a huge amount of data as they work to sense their surroundings. Some (unverified) estimates put the amount at 1GB of imagery data per second; others estimate it at 1TB every 16 minutes. The question of where to store it must be addressed.
18. We could lose as many as 50% of our truck-driver and delivery-truck jobs in future if AVs became widely adopted.

19. There are 270 million non-autonomous vehicles (cars, trucks, motorcycles, buses, trucks, etc.) in the United States alone—and about a billion in the entire world.
20. Our grandchildren may not need driver's licenses at all—with AVs available, they'll just use robo-taxis.
21. AV makers will need to address privacy concerns, as the cars will log exactly where we go and what we do while in transit. There may even be hidden cameras inside the cars, observing the passengers.
22. Fully autonomous vehicles and taxis may start first in China first because there are fewer regulations there than in North America and Europe.
23. There are several hundred companies involved in the "connected AV" sector – that will enhance communications capability of AVs. Many of these companies (such as Uber) are not yet profitable but have stock-market valuations in the billions. In the third quarter of 2017, Tesla made only 8400 cars per month, but the company's market value exceeded that of Ford, which churned out almost 100 times more vehicles.
24. The auto industry is the largest and one of the most competitive manufacturing industries in the world. It could become a three-trillion-dollar industry when smarter AVs become universal.
25. The auto industry is among the most automated industries—on average it takes only 15–20 hours to build a car from stamping to finish.

## 1.2 The Brave New World of Autonomous Vehicles

Over the last five years, autonomous vehicles, also called self-driving cars by some and "driverless" cars by others, have excited the imaginations of the media and the public alike, creating a huge platform for automotive technology visionaries, enthusiasts, evangelists and innovators. This group has painted many realistic and not-so-realistic scenarios of the arrival of these cars. By the end of 2017, early proponents had built and demonstrated, albeit in controlled environments, "semi-autonomous" capabilities in trial models of a number of cars that stood, in my estimation, at the 30–50% mark on the Society of Automotive Engineers' (SAE) calibration scale where 100% meant fully-compliant SAE level5. However, several auto manufacturers, such as Waymo (Google) and Tesla are hoping to deliver high-automation SAE level-4 vehicles (cars with an option to turn the control over to a human driver) and fully autonomous level-5 vehicles (with no option to turn the control to a human driver) by 2019 or 2021. This book is about the automotive industry's race toward safer, smarter, more autonomous and cleaner cars, whose capabilities go well beyond what's being experimented with today. That's why I use the term "AV Plus" to describe the cars of the future.

*Tesla Model S with Autopilot (Source: Tesla)*

*Google self-driving car and a blind person (Source: Waymo)*

*Figure 1.1: Driverless concept car (Source: Mercedes)*

Tesla's Model S electrical vehicle with its Autopilot upgrade, Google/Waymo's self-driving car trials in the Silicon Valley and in Phoenix, Arizona, and Uber's experiments with autonomous ride-sharing cars in Pittsburgh (home of Carnegie Mellon University's autonomous vehicle research facility), San Francisco and Arizona have been widely written about in major newspapers and across the internet. TV and press coverage has been extensive—entirely commensurate with the innovation potential that these cars could bring about. Technology conferences are being held all over the developed world to discuss and debate the future of self-driving cars. The delivery of ADAS (Advanced Driver Assistance Systems) features in luxury and semi-luxury cars over the past three years has increased the credibility of early entrants' claims. All the relevant stakeholders (academic/research institutions, new and existing auto makers, entrepreneurs, VCs, and state and federal government regulators) are involved in this discussion. It's no longer science fiction that cars will operate without human drivers—that software robots will drive us from point A to point B. Plenty of venture capital is going into start-ups, and traditional car companies are investing a lot, too. Auto manufacturers old and new are working to offer a certain level of semi-autonomous driving today and promising higher levels of autonomous capability as early as 2019 (by Tesla), with others to follow from 2021 onward. The attention that this new incarnation of vehicles is garnering is well-justified—these innovations are tremendously positive and disruptive over the long run.

I think that there's more to these innovations than the self-driving buzz. The truth is that the cars are becoming safer, smarter, connected and entertaining hubs. However, the self-driving capability is certainly one of the most exciting parts, but the drive to get rid of the internal combustion engine (ICE)—the mainstay of the automobile for over a hundred years—and replace it with electrical batteries is also accelerating, slowly but steadily. Another exciting prospect is the idea of reducing individual car ownership, as we grow accustomed to hailing shared vehicles on demand—maybe a different car every time. With all the experiments on public roads, the general public is learning to expect safer, smarter, cleaner, connected and self-driving cars (both conventional gas-powered and electrical), which I call AV Plus cars. The question is not if but when, in what form and at what pace the AV Plus cars will arrive. We are at the beginning of a 50-year evolution. There's as much known as there is speculative; as much is certain as is uncertain. I'll explore both aspects in this book.

## 1.3 Abundant Business Press Coverage

Many news stories related to autonomous cars have caught readers' attention over the last several years— perhaps more than ups and downs in the stock market, the war in Syria or the situation in North Korea:

- Uber had to abandon its AV driving trials in Tempe, Phoenix after its vehicle driving in AV mode killed a pedestrian near a cross walk in March 2018.

- A Tesla driver died in a car crash while watching a DVD and letting the Autopilot software control the car.

- Google's autonomous cars had driven five million miles in the United States without a fatal accident (although there were some fender benders) by June 2018.

- A blind man rode in a Google autonomous car in 2016 and praised the vehicle for removing a major impediment for blind people.

- Almost all major car manufacturers have announced target dates for fully autonomous (SAE 5) cars that are less than five years away.

- Many cities are eager to woo AV manufacturers by offering controlled areas and roads for testing of autonomous cars.

- At major car shows in Detroit and Barcelona, there were more curious spectators in Tesla's electric-car booth and Google's autonomous-car booth than in Maserati's sports-car booth.

- Intel bought Mobileye of Israel in a $15.3-billion deal and moved its own automotive unit to Israel. Intel has predicted that autonomous driving will result in a $7-trillion boon to the world economy by 2050.

- Volvo CEO Håkan Samuelsson has announced that all new Volvo cars designed after 2019 will be electric.

## 1.4 Why Are Automobiles so Important in Our Everyday Lives?

Since the industrial revolution, cars have had a profound impact on our lives and mobility. They affect where we live, where we work, how we interact with each other, how we move goods and how we go about our days. Cars make up perhaps the single largest expense in western households after housing, food and health needs—in terms of both capital expenditure and ongoing expenses—and they also provide the greatest utility and contribution to the twenty-first century western lifestyle.

- The auto industry is one of our society's largest employers, providing jobs across production, sales and service.

- Automobiles cause several unsolved urban problems, such as congestion, air pollution and injury/death.

- The average person in the western world's congested cities—where the desire to own a car may be the highest—spends 15–20% of their daily time in the car or on public transit, including commuting, shopping and other activities.

- Cars are the only individually owned items that take up 12–15% of our household space (a two-car garage in a modern North American house takes up 360 square feet).

- The auto industry is perhaps the only industry that has its own exclusive section in major daily newspapers—auto journalists are in a class by themselves.

Clearly, cars are an important part of our daily lives.

## 1.5 AVs or E-AVs Will Bring Disruptive Changes to Our Lives

Many industry observers and analysts feel that the introduction of fully autonomous cars will introduce extremely disruptive changes to the automobile industry, to the lives of consumers and to the delivery of public transportation services. Some feel that electric

AVs (E-AVs) might bring about the most significant change in our lives since the invention of motorized vehicle in late 1890s. While I will provide a more detailed and realistic analysis of the pace of change in chapter 14, I am providing the following forecast for the next 40 years of potential disruptions from the viewpoint of an AV optimist:

- Making cars safer, smarter and electric implies only incremental changes to our basic mode of transportation. But removing the steering wheel, accelerator and brakes and replacing the driver with an "AI-powered robo-driver" are major changes. Even the airline industry has not done that completely; although it is possible to fly a plane on autopilot, there are very few obstructions in the air. As all occupants of an AV become passengers, the notion of the licensed driver will change or disappear.

- Since several established auto original equipment manufacturers (OEMs) have announced they will produce only electric cars beyond a certain date, and others will surely make similar commitments, E-AVs will sound the death knell for internal combustion engine cars by mid-century.

- Consumers from the millennial and Y generations will buy fewer private cars. With decreasing car sales, established car companies will have to find different channels of revenue, such as TaaS, computer apps and subscription-based services in partnership with telecom companies. A McKinsey report forecasts a bullish revenue outlook for established OEMs until 2030, but from 2040 onward, if the AV revolution goes on as expected, the outlook for auto manufacturers is anybody's guess.

- Familiar gas station kiosks will be replaced by hybrid gas cum charging stations. Superfast charging time may be in the ten-minute range.

- Some oil companies (such as Exxon/Esso, Chevron, and Shell) will survive or merge; others will disappear.

- Robo-drivers will be safer and more cautious drivers. This could lead to a significant reduction in deaths and injuries due to accidents. Our health care costs due to vehicle accidents would then go down.

- Legislators and consumers will demand lower insurance premiums. The insurance companies will find it difficult to survive. They will offer UBI (usage-based insurance).

- Traffic jams will reduce eventually. This will invigorate our downtowns. There will be a notable movement of millennials to downtown condominiums in large cities.

- Only affluent sector of population will own private cars. General public will own fewer cars because TaaS will become universal at lower rate than current taxi/Uber-like services. Newer houses may not have garages.

- AV apps (modified smartphone apps and many other generic AV apps) will be integrated into car infotainment systems. Auto OEMs will derive revenue from these apps to recoup their investment and survive.

- OEMs will struggle to re-invent themselves to survive. Some may be acquired by equity-rich Silicon Valley companies like Alphabet (Google's parent company) or Apple. Their workforce would consist of robots, AI programmers and retrained robot operators.

- Cities will have to find new sources of taxation as gas tax, car registrations and infrastructure usage decrease.

- Public safety agencies will be able to reduce the number of traffic police officers and traffic-offense courts.

- Public roads will have to be reconfigured with smarter traffic signals, clearly marked lanes and smarter road signs.

- Cars will be connected wirelessly to other cars, to city infrastructure and to Internet cloud.

- Auto dealerships will be replaced by self-driving familiarization and trust-building centers. New business models will evolve to sell ride packages (along the lines of 100 zone-one rides per month for $399 or 1,000 ride miles for $500).

## 1.6 Clarifying Terminology—Self-driving, Driverless, Autonomous or Connected Cars

Let's clarify the terminology used in the media first. The media buzz is primarily on the *driverless* feature of this disruptive technology. It's obvious that driverless cars are self-driving, but self-driving cars may not be completely driverless: there may be a human driver in the car ready to take over in emergencies. Autonomous cars can be semi-autonomous or fully autonomous. We'll look at all of these incarnations of autonomous cars.

In fact, I want to go beyond the media's major emphasis on autonomy because cars are becoming smarter, more productive for occupants and more entertainment-centric as infotainment hubs. While these capabilities may seem relatively simple to implement technologically because of the mobile phone and wireless internet evolution of the past decade, there are a number of user-interface issues, industry-standards issues and mobility-integration issues (between the phone, the entertainment channels and the car's computer operating system) that increase the complexity of future autonomous cars.

I'd even suggest that the word "autonomous" doesn't fully describe what's going on or what will happen to the future incarnation of our favorite mode of transportation. It may be the central or most exciting aspect of the changes in the car but the term "autonomous" falls short of describing all the advanced features that may be introduced into future versions of high-end cars.

We don't have a jazzy and easily pronounceable acronym to describe the future car. But the term "**AV Plus**" is at least better than *Safer, Smarter, Greener, More Connected and More Info-centric Vehicles of Varying Degrees of Autonomy*. AV Plus may be powered by ICE (in the interim) or electric batteries but should have all the incremental enhancements like connectivity, infotainment, smartphone integration and a new universal user interface.

## 1.7 Levels of Autonomy (*Source – SAE Classification Table Reproduced from SAE International Website*)[1]

For some time, there was a lot of confusion in the trade press about the level of autonomy in different experimental versions of cars. Tesla was partially responsible for causing this confusion by adding its "Autopilot" terminology to its S-model car. Tesla's Model S was only partially autonomous and always expected a human driver. Google, on the other hand, extended the envelope of automation much further and started experimenting with cars that did not require a human driver. While this was going on, established OEMs (Ford, GM, Daimler, BMW, etc.) were retrofitting their cars with ADAS features and projecting future cars that would be fully autonomous. To avoid this confusion, the National Highway Traffic Safety Administration (NHTSA) tried to formalize different levels of automation based on the amount of human driver intervention expected. At the same time, the SAE, on their own, established six levels of automation and published a standard—codified as the standard: Taxonomy and Definitions for Terms Related to On-Road Motor Vehicle Automated Driving Systems. The J3016 standard "provides a harmonized classification system and supporting definitions that:

- Identify six levels of driving automation from 'no automation' to 'full automation.'

- Base definitions and levels on functional aspects of technology.

- Describe categorical distinctions for a step-wise progression through the levels.

- Are consistent with current industry practice.

- Eliminate confusion and are useful across numerous disciplines (engineering, legal, media, and public discourse).

- Educate a wider community by clarifying for each level what role (if any) drivers have in performing the dynamic driving task while a driving automation system is engaged."

Figure 1.2 summarizes the different levels of automation as identified in SAE International Standard J3016.

| SAE Level | Name | Narrative Definition | Execution of Steering and Acceleration/ Deceleration | *Monitoring of Driving Environment* | Fallback Performance of *Dynamic Driving Task* | System Capability *(Driving Modes)* |
|---|---|---|---|---|---|---|
| *Human driver* **monitors the driving** | | | | | | |
| 0 | No Automation | Full-time performance by the human driver of all aspects of the dynamic driving task, | Human driver | Human driver | Human driver | n/a |
| 1 | Driver Assistance | D*riving mode*-specific execution by a driver assistance system of steering, acceleration, deceleration using information about the driving environment, expecting that the *human driver will* perform all remaining *dynamic driving* | Human driver and system | Human driver | Human driver | Some Driving modes |
| 2 | Partial Automation | Driving mode-specific execution by driver assistance systems of both steering and acceleration/ deceleration using info about the driving environment with the expectation that the human driver perform all remaining aspects of the dynamic | System | Human driver | Human driver | Some driving modes |
| *Automated Driving Assistance System* ("system") **monitors the driving environment** | | | | | | |
| 3 | Conditional Automation | D*riving mode*-specific performance by an *automated driving system* of all aspects of the dynamic driving task expecting that the *human driver* will intervene if required. | System | System | Human driver | Some driving modes |
| 4 | High Automation | D*riving mode*-specific performance by an automated driving system of all aspects of the *dynamic driving task*, even if a *human driver* does not respond to a *request to intervene.* | System | System | System | Some driving modes |

| SAE Level | Name | Narrative Definition | Execution of Steering and Acceleration/ Deceleration | Monitoring of Driving Environment | Fallback Performance of Dynamic Driving Task | System Capability (Driving Modes) |
|---|---|---|---|---|---|---|
| 5 | Full Automation | Full-time performance by an *automated driving system* of all aspects of the *dynamic driving task* under all roadway and environmental conditions that can be managed by a *human driver*. | System | System | System | All driving modes |

*Figure 1.2: Summary of Automation Levels - Standard J3016 (Source: SAE International)*

In 2016, the NHTSA in the United States adopted SAE Standard J3016 and is in the process of putting more flesh on these specifications, including testing and certification criteria and infrastructure. German automobile regulators are moving quickly and in sync with the NHTSA; similar bodies in other countries will follow the NHTSA's lead.

## 1.8 Possible Future Scenarios

Let's look at some of the ways people's lives may be affected by AV Plus cars over the next 10 to 30 years.

### Scenario 1—Urban Family Adopts AV Plus Enthusiastically
An upwardly mobile family (Carla and Dan Potter) used to own two conventional ICE cars in Austin, Texas; one for Dan, one for Carla. Now they lease only one self-driving (fully autonomous, level-5) car. The car drops the Potter children, Charlie and Elizabeth, at the local school and then takes Dan to the office by 9:00 a.m. Carla uses the car for work for most of the day, then sends it to pick the children up from school at 3:30 p.m. and Dan from work at 5:30. The family also uses Uber or Lyft for those occasions when they need an extra car. On the weekends, the family rides in the AV as if they're in an expensive limousine. They buy insurance coverage under UBI (usage-based insurance) from Geico. As a result of switching from a two-car family to a "one-car-plus-TaaS" family, the Potters have reduced their capital outlay in cars, reduced their monthly operating expenses including their insurance cost, and reduced their carbon footprint. If the cost of TaaS continues to go down, they may consider not renewing the lease on their AV and instead use TaaS providers for all their transportation needs.

### Scenario 2—AVs Deliver Pizza and Amazon Prime Goods
Sylvia lives in a gated community in Ruby Hills in the Bay area of California. She orders a pizza from the local Domino Pizza franchise through his Alexa system. The Domino Pizza store prepares the pizza and sends its unmanned Ford delivery van to Samir's area, along with pizza deliveries to adjacent areas. A drone mounted in the van loads the pizza from a tray, opens a hatch in the roof and drops the pizza at Sylvia's door. Pizza delivery app informs Sylvia by a text message that Pizza is there at the entrance and the drone returns to the van. The van drives back autonomously. The app sends a reminder text message

to Sylvia if he does not acknowledge the previous message. No pizza delivery driver is involved. Interestingly, Ford and Domino are already conducting experiments like this in New Zealand.

*Figure 1.4: Ford and Domino Pizza Delivery in New Zealand (Source: Wired Magazine)*

## Scenario 3—Business Executive Uses Travel Time for Client Meeting

With his robo-chauffeur doing the driving, Pat Smith, a busy business executive in Australia, loses no time due to travel. Today, he's making a marketing pitch to a couple about a franchise deal, using the travel time to the prospective site to talk business instead of having to focus on driving.

*Figure 1.5: A Business Meeting in an SAE Level-5 AV Plus – Source Mercedes*

## Scenario 4—Sales Leader Makes Productive Use of His Commuting Time

Lois Richard is vice president of sales at an insurance company in Paris, France. Lois lives in a high-rise condo building and owns a level-5 Audi 7 autonomous car because he drives almost 50,000 miles every year for business. Lois used to spend two hours a day driving around the city. Now, before leaving the condo in the morning he uses his iPhone to call his Audi, which is parked in his condominium garage, and the car is waiting for him when he gets downstairs. His schedule is already in the Audi's computer because it's synchronized with his phone. The car drives him to his appointments while Lois catches up on all his emails and text messages, making valuable use of his commuting time. He leaves the car outside when he goes in to each appointment and the Audi parks itself.

## 1.9 The AV Plus Opportunity is Big—and the Impact on Our Lifestyle is Even Bigger

The auto industry is perhaps the third or fourth largest industry when measured by revenue generated, GDP added, portion of our disposable income spent and workers employed. In raw dollars, the transportation industry based on passenger cars, commercial autos and trucks is enormous. According to Statista, a statistics publishing portal, the size of the auto OEM industry is around US$1.5 trillion. According to 2017 figures, the conventional automotive industry (made up by Toyota, Volkswagen, GM, Ford, etc.) produces 75–80 million cars every year, and there are over a billion cars on the road around the world. If we add parts and associated services, it could be a US$2.5- to $3-trillion industry. Purely in terms of manufacturing, it's probably the largest unified industry. In fact, the auto industry is bigger than the information technology industry, the telecom industry and the pharmaceutical industry. In terms of employment, according to national industry association reports as documented in a 2015 McKinsey report, the auto industry (including services) employs over 12 million people in Europe, six million people in the United States and around five million people in Japan. The average North American household spends $12,000 to $15,000 each year to maintain and operate two cars. That's quite a bite from our monthly pay checks.

The disruptive changes coming to the auto industry and to consumers through AV Plus innovations in the decades to come will clearly have a significant impact on our lives. These changes will also impact the overall employment situation in the industry, our surrounding infrastructure, especially in congested cities, and the environment.

## 1.10 Auto Industry was Passive with Self-driving Innovation till Google Showed Up

The auto industry is over a hundred years old. In the United States, Ford launched its Detroit operations in 1899, and in Germany, Daimler was established in 1909. The Japanese auto industry came into prominence as a major exporter after the 1950s. For the past hundred years, the industry has been making incremental improvements in the way cars are designed and the way components are assembled and integrated into a functional car. We have seen major improvements to how cars are operated through inventions like automatic transmission (introduced in 1940 by GM) and cruise control (which was invented in 1948 but became common in the 1960s). Seat belts and air bags made cars safer in the 1970s. Many incremental changes were made in the 1980s and 1990s, but none were particularly significant, despite the fact that researchers in Europe and the United States were experimenting with components that would enable self-driving cars. In 1987, Europe's Eureka PROMETHEUS project made significant advances and the 2005 American DARPA project threw out a challenge to American innovators to show off their stuff in competition with Europeans. Meanwhile a small Israeli company called Mobileye was creating smart devices that would make cars safer with ADAS. In 2009, Google stole the show and put all of those efforts and discoveries into a credible self-driving project. Established auto industry that had been passive all along woke up. Detroit had been passive not taken notice for several years.

## 1.11 Computers Invade Autos and then Take Over

Computers penetrated the car industry in the 1960s. The traditional car was and is bigger in mass than all the computers it contains. Computers are small but expensive components—hidden somewhere in the engine compartment. The first major use of a computer in a car was for ignition control in the engine. In 1968, Volkswagen introduced the first computer-controlled electronic fuel injection (EFI) system, manufactured by Bosch. ECU (engine control unit) is a powerful and specialized computer processor within EFI. It collects a lot of data from engine sensors and uses that data for efficient firing of spark plugs and optimal timing of opening the fuel injector. This became necessary in order to meet government's fuel-efficiency standards for lowest emissions and highest mileage in a car.

Soon after Volkswagen's introduction, more and more industrial computers were embedded in cars' mechanisms for monitoring, controlling and activating mechanical components like electronic fuel brakes, steering, transmission and even suspension. Today, a car can have well over 60–75 computer systems, monitoring and controlling everything from ride handling to onboard entertainment and communication systems. Now with AV Plus, every car will feature a very powerful supercomputer with enormous storage capacity for storing all the computer-vision data of the car's surroundings.

The use of industrial computer chips in cars was a gradual and incremental effort by the auto industry itself. But three major events happened that would turbo-charge the race to autonomous vehicles: 1) Elon Musk's founding of Tesla, 2) Google founder Sergey Brin's self-driving car project and 3) Mobileye's introduction of increasingly more capable ADAS products that allowed established auto OEMs to begin taking steps toward autonomous cars.

Musk wanted to use as much computer technology in the architecture of his electric car as he could: he wanted to use hardware robots to assemble the cars and software robots to drive them. Tesla's Model S caught the imagination of forward-looking Silicon Valley–type consumers so much so that he had no problem raising money for his venture and expansion. He built a product that consumers loved, launching in 2012. Stock-market support for Tesla and demand for the Tesla S-model convinced the established OEMs that both the electric car and the autonomous car phenomenon were for real.

Sometime in 2009, in Palo Alto, just a few miles away from Tesla's Fremont, California, headquarters, another young technology freak, Sergey Brin, one of the two founders of

Google, started the company's self-driving car project seriously under the leadership of Sebastian Thrun. Unlike Tesla, this team did not build a brand-new car—instead they added some computer gear (sensors and more; see Chapter 3 and 7) to a Toyota Prius and built smart software that could mimic a human driver. Sergey was less interested in the electric aspect of the future car as he was in making it self-driving. As Google built AV prototypes and conducted trials, first on a private Google campus and then on selected public roads, the company continuously refined its hardware and software. Thus an ordinary Toyota Prius became an early version of Google's self-driving or autonomous car. Google shared the results of its trials with outsiders, and the media took note: this, too, was for real.

The established auto industry saw what was going on in the Silicon Valley as well as in Europe, where similar efforts were going on, and was forced to join the race. Established OEMs realized quickly that their future could be threatened by the likes of Tesla, Waymo (the new name given to the company leading Google's autonomous car efforts; now an Alphabet subsidiary) and ride-hailing company Uber who has ambitions to use self-driving cars for its ride haling service in future. Thus the race to build the autonomous car began earnestly in 2009. And so did the rivalry between Detroit/Munich/Stuttgart and Silicon Valley.

Meanwhile, a start-up in Israel by the name of Mobileye was doing some interesting R&D to produce automated crash-avoidance products that would make driving safer, building on the preliminary work done in Europe in the 1980s. Mobileye started supplying those components to auto OEMs, who began installing them in luxury and semi-luxury brands. These components were subsequently categorized as ADAS features. Mobileye became a leader in this area and, by the end of 2017, its products had been installed in 15 million cars. Mobileye's niche expertise became so noticeable that Intel acquired it in 2016 for $15.3 billion—the largest AV-related acquisition to date. The auto OEMs noticed the deal and the valuation of AV start-ups shot up.

## 1.12 Consumers Want Driver Assistance First, Integration with Smartphone Second ... and Self-Driving the End Objective

Most of the media buzz is about the end objective of the driverless cars. The idea of driverless driving is incredibly exciting, a game-changer that defies conventional thinking. It's a disruptive move for the industry and will completely change the way we drive. Look, George – no hands, no feet! Just some voice commands! People want to know: is artificial intelligence better than real human intelligence? Add to this the newly minted concepts of deep learning and computer vision—it becomes movie material. A quick scroll through Netflix clearly shows that AI has captured Hollywood's attention.

But surveys of today's consumers indicate they still don't have enough trust in driverless cars—instead they want greater safety and accident-avoidance features first. Then they want more advanced driver-assistance features. For the past few years, state and provincial motor vehicle regulators have been ticketing drivers for distraction while using their smartphones. Yet consumers want to be able to want to use their mobile devices safely in their cars. They want Apple CarPlay and Android Auto on built-in touch display in the car. Soon enough, having built trust in their ADAS, consumers will be willing to let AI-based software chauffeurs (call them robo-chauffeurs) do the driving while they relax—have a drink, watch a movie. Soon enough, the personal car will become a luxury limousine at a cost that many more can afford. This will happen one day (like others, I have my own speculations about when; see Chapter 14)—it'll take time, of course, and it may take several incremental steps over a long stretch, but it'll happen.

In Chapter 14, I'll discuss the pros and cons of a slower, natural path of progression versus a more immediate, disruptive path where fully autonomous cars enter dealer showrooms in 2021 or 2025 or 2030. Limited disruption in the coming decade will accelerate the natural progression toward faster adoption by early millennial adopters and wide adoption by Generation Z.

## 1.13 TaaS (Transportation as a Service) Will Replace Ownership, but Slowly

There has been a lot of discussion about the TaaS or MaaS (Mobility as a Service) notion of ride-hailing AVs. The idea is that our "fad" of car ownership will fade away or even disappear once we can remove the driver from the car and hire AVs on a per-ride or bundle-of-rides basis. This would certainly make it much more convenient and cheaper to move people from one place to another. My opinion is that the TaaS/MaaS phenomenon will likely replace the taxi industry but that it won't replace private ownership as extensively and as fast as its promoters and enthusiasts suggest, because the business model they're using is flawed. See Chapter 8 for my analysis of the TaaS story.

## 1.14 Technology Transformation and Adoption are Complex Processes and Generation-dependent

In Chapter 2, I'll talk about the complexity of technology transformation. Many technologies get introduced—some succeed, others fail. There are inherent obstacles and inertia of the past to be overcome. Looking back at previous technology-adoption efforts can provide some guidance, although the rate of change is increasing and adoption of disruptive technologies becomes easier and faster with every new generation. Also, I will show in Chapter 2 that technology adoption is highly dependent on generational attitudes and financial capacity. It's also important to understand that discretionary spending power generally rests with older generations. Someone can scrape together enough cash from their piggy bank and part-time job to buy a new shiny smartphone, but it's not so easy to do the same for an autonomous car that is going to cost a lot unless TaaS model takes off.

## 1.15 There Are Many Issues and Perspectives in This Space

In a major technology innovation like autonomous vehicles, there are many stakeholders and they all have their own unique issues and perspectives. Not all of these perspectives converge to the same ultimate destination in the same timeframe, and each stakeholder has an impact on the other. The path the industry follows will be a complex function of the power and capability of the stakeholders conditioned by the preference of customers who will decide which horse to bet on. In Chapters 9 and 10, I'll discuss the perspectives of the following stakeholders:

- Academics and researchers
- Developers and implementers
- Innovators
- Established OEMs and suppliers
- Infrastructure owners
- Silicon Valley players (like Intel and Nvidia)
- Software integrators
- Market promoters
- Customers from different generations
- Insurance (legal and ethical)
- VCs and investors
- Regulators (Department of Transport, NHTSA and similar bodies around the world)

In Chapter 11, I'll propose a methodology for a holistic, composite viewpoint that regulators should consider to guide the industry. The industry needs this direction based as much on consumers' ability to adapt to the new environment as on the industry's rather optimistic forecast of reduced accidents and congestion.

## 1.16 Industry Trends Impacting Future Car Architecture

In this book, I'll talk about the shape of future car architecture. There are a number of auto industry trends that are influencing the future car architecture:

I. **Emphasis on Reducing Carbon Emissions**
Consumers' fascination with Tesla's electric car suggests that the internal combustion engine will have to cede its dominant position to electric cars so long as the barriers (artificial and real) to this switchover are removed.

II. **Complexion of Cars Changing with Greater Use of Computers, Electronics and AI**
The car will no longer be a purely mechanical machine with an internal combustion engine, an automatic transmission, rubber tires and a steel/plastic chassis. It will become a combination of steel (along with other metals and plastic) and silicon with sensors, cameras and radar equipment. I'll discuss how the architecture and design of AV Plus cars will be influenced more by Silicon Valley automation visionaries than by Detroit/Munich/Stuttgart mechanical engineers.

III. **Road Infrastructure will Change with Smart Signs and Road Markings**
AV Plus will work best if it is supported by intelligent road infrastructure that includes smart signs that can communicate with cars and send traffic measurement information in real time to the computers that control traffic signs.

IV. **Driving is Boring … but Our Love for 24x7 Smartphones is Going Strong**
AV enthusiasts are saying that driving is a boring chore. I'm not sure whether that's true for everyone, but it does appear that driving isn't seen by the younger generations as a fun activity the way it might have been by their parents or grandparents. They want to be constantly connected through smartphones and virtual reality, and driving can interfere with that. It may be these younger generations may adopt AVs for what they do best—transport people and goods from point A to point B.

## 1.17 Unified Operating System and Integrated User Interface

We are so engrossed with AI and automation in AVs that there is only a limited amount of discussion in the academic circles, media and development labs about the need for a unified operating system environment and a consistent, integrated user interface with common elements and conventions across the industry. Every vendor is going on its own proprietary path. This approach will slow the adoption of AVs. I'll devote a chapter to this topic—why it's important, what the framework should look like and what attributes it should have (taking some cues from the smartphone industry).

## 1.18 This Book—A Primer for Interested AV Professionals Seeking Broader Perspective

This book is for business professionals and serious consumers who are looking for introductory yet comprehensive information on autonomous cars—not just self-driving cars but the cars of the future that will be safer than today's cars, smarter than today's cars, greener than today's cars and more connected than today's cars. These connected cars will not only communicate with other cars in the vicinity but also with the cloud to access information and entertainment content through a unified user interface such that it won't matter whether you use the smartphone or the screen in your car.

This book offers:
- A comprehensive discussion of AV topics from (some) technical and business perspectives.
- A high-level explanation of the relevant technologies, including AI.
- An independent critical analysis and viewpoint.
- A consultative analytical approach.
- A realistic assessment of the timing of the arrival of the ultimate AV Plus for mass adoption.

- An independent consultant's insight into optimal strategy for managing the AV revolution.

## Summary

*In this introductory chapter, I've set the stage for the autonomous vehicle scene filtered through my own lens and influenced by others and my experience with emerging technologies. I've introduced the term AV Plus, which goes beyond self-driving to include electric batteries as fuel, safety as a core requirement, all-compassing connectivity for communication, home-like infotainment, intelligence embedded everywhere and self-driving as a given.*

*AV Plus will ultimately transform our world of transportation, but not so soon. This book is a critical assessment of the state of the industry. I'll analyze how emerging technologies transform societies, suggesting that disruptive changes of this nature often face resistance from older generations, where trust in new technology can be shaky. I'll also emphasize that technology goes through significant rework and upgrades as the rubber hits the road—pun intended. Read on to learn about the fundamentals of technology adoption, the rationale for AV and the history of the AV industry, as well as some of the technology components, consumer perspectives and challenges—plus my analysis of how the AV Plus industry will evolve. In the last chapter, I'll provide some common-sense recommendations for the way forward.*

---

**Citations for External References**

[1] SAE Levels of Autonomy https://www.sae.org/standards/content/j3016_201401/

# Chapter 2

# Fundamentals of Technology Innovation and Adoption

*I learned a great lesson during my undergraduate degree at IIT (a world-class technology institution in India). I was taught that whenever you have to understand a new concept, whether it's technological, engineering-related, business-oriented, economical factor or human-factor related, you should ask yourself: "what are the fundamentals of this concept?" The rationale for this approach was simple: fundamentals govern a phenomenon and its outcome more than incidental factors.*

*In my view, the AV conversation has a few fundamentals that will govern its success. This is indeed a complex conversation because AV Plus is a multi-disciplinary issue. It may, on the surface, appear to be a discussion about technological innovation and business models. But that's not all that's involved. It may appear to be about a revolution/evolution of the auto industry—it's not only that, either. It may be about the excitement of self-driving cars—it is, but it's more than that, too. It may appear to be about a paradigm change from car ownership to transportation as a service; it may involve the promise of reducing traffic accidents and auto-related deaths; it may be about trust, ethics and habits. The AV discussion involves all of these elements, and many more. I'll touch on some of the fundamentals in this chapter and throughout the book.*

**Media Coverage of AVs Lacks Adequate Discussion of the Fundamental Issues**

There's a lot of discussion about predicting how fast AV technology is coming down the road. In articles by business journalists and industry innovators, including academia, analysis of various factors determining the pace of adoption is relatively simplistic—it is primarily based on the enthusiasm of the AV technology startups and Silicon Valley challengers to the established OEMs. Some of the consulting firms like McKinsey, Accenture, Deloitte and others have done a better job of discussing the inherent challenges the industry faces and of forecasting a more realistic picture, and yet their published reports are at a somewhat high level (since they're designed to attraction client attention in the hopes of being engaged for more serious, paid, analytical work). As a result, there is scant discussion of the fundamental factors affecting the technology-adoption cycle in the context of autonomous vehicles. Quite often, best-case scenarios are presented using the most optimistic assumptions.

This lack of deeper analysis of fundamental factors is due to the fact that much of the content is written by busy business journalists, observers and bloggers who write "easy-to-understand" stories for consumers. It's exciting to write that a new scientific discovery has been made or that a research lab has made a technological breakthrough—those stories sell. Quite often, journalists craving newsworthy material seek the information from early pioneers and developers of new technologies, and the stories they then write lack in-depth analysis of the assumptions made by the promoters of new technology.

We can't blame business writers. It's less interesting to say that early technology innovators may face many unknowns. Everyday readers don't have the time or the patience to read in-depth analysis. They just want to know if, when, where and what—mainly the features, the likely price range and when it'll be available. Business writers are simply the messengers.

**This book is a bit different.** Here, I have the space to discuss the fundamentals of AV Plus innovation and adoption. I spent many years as a technology consultant with a focus on the critical analysis of technology adoption. I also have real experience with emerging technologies (and have the battle scars to show for these struggles). So a bit of skepticism is ingrained in me. I believe that what we really need when it comes to understanding the evolution of AV is an in-depth reality check—a deep discussion of the fundamentals involved in the AV Plus story. That's what this book is about.

## 2.1 Fundamental #1—Technological Innovation and Adoption Involve Many Twists, Turns and Reworks

Much has been written in academic papers about the technological innovation and adoption process, and a lot of research has been done into the inherent revision and rework involved in taking a product to market ... but it's often not given enough attention in the press. Consumers don't have an appetite for it and proponents have no motivation to talk about it. But it's important to acknowledge the many, many uncertainties, turnarounds, diversions and revisions that new technologies go through before consumers adopt a solution *en masse*. I'm not talking about technology enthusiasts or early adopters who will try anything new the first moment it appears. I am talking about the population at large and mass adoption.

Typically an idea is conceived in the mind of a scientist or researcher in an academic institution. Sometimes a business innovator goes to a scientist and asks him/her whether it's possible to produce a product to meet a certain need that they've identified. Often the idea is only conceptual in nature, and if it's ground-breaking, it generally requires a lot of research—both theoretical and experimental. Universities keep extending the boundaries of knowledge, testing hypotheses and theories. Research results get peer-reviewed and published. Ideas get patented. We'll call this **stage one**.

In a subsequent stage (let's call it **stage 2**), industrial research labs or product development teams in industrial organizations start investigating whether there's a business opportunity for converting the idea or research into a marketable product. Quite often, it's not the development lab that evaluates the business opportunity but an enterprising innovator. The entrepreneurial team, consisting of a scientist and a business person, starts developing a prototype. They start showing *alpha* prototypes to VCs and angel investors. With their VC money, the team grows and develops a more advanced *beta* prototype. They start validating the idea and testing their beta prototype in an internal (synthetic) user environment.

Stage 2 is followed by **stage 3:** actual customer trials. This is the real test, conducted by early adopters who are keen to try out a brand-new, shiny product, often to make a statement or to show that they're ahead of the crowd. The early adopters may find some things right but other things wrong with the prototype. These enthusiastic customers encourage the innovator to keep improving the product. The engineer and scientist go back to the drawing board—the lab or the academic research environment. The product gets redesigned, reassembled and relaunched. After several iterations, the product either is abandoned or goes into a controlled field trial—where the rubber hits the road.

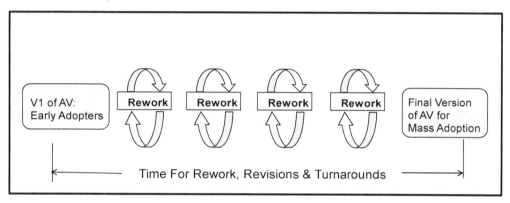

*Figure 2.1: Rework as an Inherent Phenomenon in Technology Evolution*

The key point is that there is an inherent "rework" loop everywhere till the product reaches a steady state in which it's ready for mass adoption. This rework takes time and capital. For complex products like AVs that have many different elements controlling adoption, the process can take years or decades, especially if a breakthrough in some technology component is required and it's still in the lab. **The more complex the technology-integration task, the more frequent and longer the rework loops may be**.

## 2.2 Fundamental #2— Generational Gap Affects Technology Adoption

When a technology is disruptive and requires a significant change in consumer behavior, the attitudes and opinions of different generations are extremely important. While there has been some discussion on this subject in various articles and research papers, I'll explain a few fundamentals here:

1. The younger we are, the less we are governed by habit and the more likely we are to adopt new technologies. Resistance to change is directly proportional to age. As a child, we are like sponges, absorbing change without any resistance. The higher our age, the higher our resistance to change.
2. The younger we are, the more important it is for us to make a statement or fit in. Almost every kid wants the latest and greatest brand of shoe or smartphone that they see on TV.

3. The younger we are, the more trust we have in new technology and less concerns we have about reliability and safety.
4. Every generation spans approximately 25 years and generational time period affects adoption of technologies that are dependent on the attitudes of a generation..
5. When our physical mobility is a challenge (e.g., if we are old, blind or disabled), we will accept any new technology that gives us this basic mobility.

The adoption of disruptive technologies is highly dependent upon the habits of a person or group of people as they grow up. As we get older, it becomes increasingly difficult to give up our habits. Therefore, with every new generation, humans become newly receptive to disruptive technologies.

In current generation-related terminology, we divide our population into five generations—traditionalists (born after World War I), baby boomers (born after World War II), Generation X (born 1965–1980), Generation Y or millennials (1981–1998) and Generation Z (born after 1999). The attitude of these generations to technology in general and AVs in particular are different and will affect adoption rate of AVs. I am using a table developed by WMFC (West Midland Family Center) as a basis.

Table 2.1 describes some of the generational characteristics and the differences in attitudes toward technology adoption.

|  | Traditional Generation | Baby Boomers (BB) | Generation X (Children of BB) | Generation Y (Millennials} | Generation Z |
|---|---|---|---|---|---|
| Age group | Born during or after World War I (1915–1944) | Born after World War II (1945–64) | Born 1965–80 | Born 1981–98 | Born after 1999 |
| 2016 US population ( 326 million) | 32 million (9.9%) | 75 million (23%) | 65 million (20%) | 80 million (24.5%) | 74 million (22.6%) |
| Influencers and mindset | WWII, Korean War, Great Depression Hard workers | Civil War, Cold War, space travel Materialistic and ambitious | Energy crisis, dual income families, activism Take care of themselves | 911 terrorist attacks, school shootings. Want to change a lot of things | Idealists, caring, environment-conscious |

|  | Traditional Generation | Baby Boomers (BB) | Generation X (Children of BB) | Generation Y (Millennials) | Generation Z |
|---|---|---|---|---|---|
| Wealth and capacity to spend | Did not have a lot of wealth | Earned more than other generations; Have capacity to spend | Made less than BB; less capacity to buy expensive items | Less than Gen X | Still not ready to buy AVs; Outlook for AVs positive |
| Attitude to technology | Do not like too much technology | Have accepted the technology revolution | Love technology, love smartphones | Attached to smartphones and social media | Born with it – adopt it naturally |
| Attitude to driving | Loved their cars | Love their cars | Car only a necessity | Do not like driving as much | Will embrace TaaS |
| Attitude toward AV | Do not care | Do not trust AVs | Will consider AVs | Would love AVs | AVs are the way to go |

*Table 2.1: Generational differences affecting technology adoption- source www.wmfc.org*

## 2.3 Fundamental #3—New Generations' Decreasing Interest in Driving

A study conducted by the University of Michigan Transportation Research Institute found that there's a decreasing interest in driving among millennials and Generation Z (see Figure 2.2)[2]. Driving is no longer seen as a fun thing, and teenagers aren't rushing to get their driver's licenses when they turn 16. While it's often a necessity, driving is seen as rather boring. These younger generations are turning to shared transportation services like Uber and Lyft, having fewer cars per family and living in urban areas near public transit.

*Figure 2.2: Recent decreases in the proportion of people with a driver's license across age groups (Source: Michigan University Transportation Institute)*

The use of smartphones, on the other hand, has taken off among the younger generations. The smartphone is an absolute necessity for millennials and later generations, since social-media-based communication is among their highest priorities. In fact, given a choice, younger generations would generally rather have a smartphone than a car. Freedom to them means communication via social media, anywhere, 24/7. Through virtual apps, millennials can be with their friends while they're at work. They can see, watch, converse and work from anywhere through their smartphones. Physical mobility through transportation is secondary.

These differences in attitude toward driving and cars as lifestyle possessions will have a direct bearing on the adoption of AVs, particularly as AVs become available as a less inexpensive, on-demand transportation service. Younger generations will not necessarily want to have AVs of their own in their garage.

### Transportation—An Evolving Human Necessity

It is apparent from the "Generational Difference Chart" in Table 2.2 that baby boomers were very much material-asset focused. For them, owning a house and a car was important and is still important. They determined their success by the square footage of their houses, the streets they lived on and the cars they drove. These factors are much less important for the majority of millennials. Digital technology was thrust upon baby boomers, and they accepted it and adopted it. But for millennials, digital technology, the internet and social media are a symbol of and a mechanism for achieving freedom. Status differences are less important, and democratization through the internet and the global viewpoint are more desirable. Cars are often a necessity for them, if jobs are not available where they live. But if they could, many millennials would work remotely and deliver services electronically.

## 2.4 Fundamental #4—Trust is Essential in the Adoption of Any New Technology

We also need to consider the role of trust in the adoption of new technologies. Trust is extremely important. This is true for all generations, although younger generations tend to be willing to take more risks in general.

It's important to note that the main argument for AVs is about preventing accidents and reducing or even eliminating casualties due to vehicle accidents. But if AV autopilot software is going to replace human drivers, we need to be certain that it'll work reliably. When we're dealing with questions of life and death, all generations will have to be able to trust that what AV visionaries, evangelists and promoters are saying is true. Pure public relations efforts won't work—trust has to be earned not only through demonstrations and trials but through real customer experience. This will take time: much more for baby boomers, much less for millennials … and perhaps none at all for Generation Z.

## 2.5 Fundamental #5—Public Institutions Move More Slowly than Private Institutions

The auto industry is very large and also highly regulated. It is intrinsically dependent on the public infrastructure that cities and municipalities provide. State and provincial DOTs (departments of transportation) regulate the industry through licensing of vehicles and drivers, federal agencies provide overall direction, public safety agencies like the police monitor vehicle operation on the roads and the insurance industry provides legal liability protection for loss of property and life. Not all of these players move at the same speed. The reality is that public institutions and regulators move at a much slower pace than those who benefit directly from new products. It will take a long time for real regulations (and associated penalties for infractions) to be put in place. Baby boomers still influence legislators and regulators more than any other generation.

## 2.6 Fundamental #6—It's Easy to Develop a Habit, More Difficult to Change a Habit and Very Difficult to Break a Habit

Driving is a habit (or a skill) that human drivers acquire quickly because it satisfies the basic human need to move from point A to point B (and is faster than walking or riding a horse). We get basic training from driving schools, but it's up to our human intelligence to figure out how to drive in many different situations. We use our "real" intelligence when we run into situations that aren't described in the driver's handbook. We learn from each situation and log that information in an intelligent driving database in our brains. It becomes an ingrained habit.

However, switching from human driving to being driven by an autopilot will mean giving up that control, which we may find difficult, especially if we don't trust the autopilot. In most cases, AV is safer than a human driver. But are we willing to give up our perception of control?

## 2.7 Fundamental #7—The Reliability of Complex Hardware is Inversely Proportional to the Number of Distinct Components and Suppliers

Why is it that Apple products are more reliable and easier to use than others? There are several reasons. However one of these reasons is that almost every component of Apple products is designed by Apple and there are relatively few component suppliers. The same is not generally true for Android-based phones. In AVs, there are many layers of designers and component suppliers. OEMs are simply systems integrators or assemblers of components. This makes for a complex situation that will take a lot of work to get it right.

## 2.8 Fundamental #8—The Reliability of Computer Software is Inversely Proportional to the Number of Lines of Code

A modern AV contains 60–80 computers and an estimated 100 million lines of code that will need to be updated constantly, certainly in the early days of AV development. This much code is more than what was employed in the Apollo moon shot, which had hundreds of experts monitoring it and ready to intervene in real-time at any failure in the system.

## 2.9 Fundamental #9—Society's Tolerance to Accept Human Deaths as a Price to Perfect a New Technology Has Gone Down Since 1886 – the Onset of Automobiles

In the developed countries, value of human life is becoming higher and higher. While we forgive human beings for making mistakes, accidents and deaths, we expect almost faultless (nearly 100%) performance from the machines we create, such as AVs. We accept accidental deaths due to imperfect human actions, but not from imperfect autopilot software.

## 2.10 Fundamental #10—The Majority of Computer Programming Efforts are About Handling Abnormal Situations

Generally in a software development project, the programming of the basic functionality takes up only a small amount of effort. It's the huge number of potential abnormal situations and error-handling situations that take up the majority of the development and testing effort. As Professor Karl Iagnemma of MIT, the CEO of AV software startup nuTonomy, says, "the last 1% is harder than the first 99%." So, we'll need to sit tight for a long while as we wait for a trouble-free AV ride.

You will notice in later chapters that I will use these fundamentals to validate my conclusion that it will take longer for AV adoption by consumers and reasons why scientists and technology companies will have to work much harder to find acceptable levels of quality that consumers will accept.

### Summary

*Technology innovations can introduce solutions to human problems or replace existing solutions, but all of them must deal with fundamental factors that surround the adoption of technological solutions. These fundamentals span technological, psychological, generational and economic factors.*

---

**Citations for External References**

[2] University of Michigan Transportation Research Institute Study –
http://www.umtri.umich.edu/what-were-doing/news/more-americans-all-ages-spurning-drivers-licenses

# Chapter 3

# AV History—How We Got Here

*The idea of a self-driving car is not new. Many attempts at building one have been made over the past century. From 1920 to 1960, several experiments were conducted to remotely control cars through radio waves first and then through embedded electronics in roads on protected race course-like tracks. But motivations were different then. The main goal was not to save deaths due to car accidents; it was the excitement of autonomy and convenience. Automobile engineers developed cruise control first, and then adaptive cruise control, automated braking and other advanced driver assistance systems. The introduction of computers assisted in the driving task. The European Eureka PROMETHEUS project between 1987 and 1995 and the American DARPA challenge in 2004–2007 helped to spur innovation, and universities like Bundeswehr (Munich), Carnegie Mellon University Pittsburgh), MIT and Stanford began to research autonomous cars in a serious way, with a lot of funding and government support. The technology got a major boost when Google and Tesla got into the action in 2009. In 2012, Professor Geoffrey Hinton's research into artificial intelligence, at the University of Toronto, began to play a very important role in this journey. As we race toward the realization of the future autonomous car, it's important to remember its history.*

Automating the driving function in a car has been an elusive desire of enterprising automotive engineers and avant-garde drivers alike for a long time. In fact, in the early days of the automobile, many observed that most accidents took place because of human error, and automotive engineers have been looking at various ways to assist the driver ever since Henry Ford started mass-producing cars. Fast forward to today, and engineers are working to take the driver out of the front seat all together. Let's look briefly at some of the notable attempts in this effort over the years.

## 3.1 Driverless Car Efforts in the Twentieth Century

Early auto technology innovators looked at radio waves as a means to control cars remotely. Radio engineers knew that they could transmit radio signals from one place to another. They believed that as long as the signals could convey the necessary information, they could create a device that would use that information to control the steering and throttle of the car. This was the basis of the first few attempts at self-driving cars. Interestingly, electric cars were in competition with gas-powered cars back then, before gas ultimately won out.

### 3.1.1 The First Attempt at a Driverless Car: 1925

In 1925, Francis Houdina (who owned Houdina Radio Control in New York) caused a stir in the industrial world by driving a remote-control car up Broadway and down Fifth Avenue, using radio waves. As you can see in the photo of Figure 3.1, the car he used was a retrofitted 1926 Chandler—an open-carriage car called the American Wonder. His setup was a two-car combination—the first car was "self-driving," and it pulled the second car, which contained the radio, behind it. The first car's antenna picked up radio waves, which were used to control its transmission shaft, with no human driver. Of course, Houdina did have a backup driver, in case things didn't work out.

*Figure 3.1: 1925 Self-driving demonstration in New York (Source: Wikimedia Commons)*

The demonstration itself was kind of shaky and rough. According to the *New York Times*, the car almost hit a couple of nearby trucks (which ultimately drove to safety). But it did demonstrate the basic concept. Figure 3.2 shows an excerpt from the *NYT* archives.

> A loose housing around the shaft to the steering wheel in the radio car caused the uncertain course as the procession got under way. As John Alexander of the Houdina Company, riding in the second car, applied the radio waves, the directing apparatus attached to the shaft in the other automobile failed to grasp it properly.
>
> As a result the radio car careened from left to right, down Broadway, around Columbus Circle, and south on Fifth Avenue, almost running down two trucks and a milk wagon, which took to the curbs for safety. At Forty-seventh Street Houdina lunged for the steering wheel but could not prevent the car from crashing into the fender of an automobile filled with camera men. It was at Forty-third Street that a crash into a fire engine was barely averted. The police advised Houdina to postpone his experiments, but after the car had been driven up Broadway, it was once more operated by radio along Central Park drives.

*Fig 3.2: Excerpt from an article about the first self-driving car demonstration (Source: NYT archives)*

Achen Motor, a Milwaukee car distributor, demonstrated Francis Houdina's invention in December 1926, under the name "Phantom Auto," on the streets of Milwaukee. Then not much else happened for a decade.

## 3.1.2 From Radio Control to Electronic Embedding of Roadways: the 1930s

From Houdina's radio-controlled efforts, inventors moved on to a new idea: embedding electronics underneath the roads. A decade after Houdina's demonstration, Norman Geddes, a transportation visionary, talked about his vision of the *City of the Future* in which highways would be embedded with electronics. In Geddes's Futurama exhibit (Figure 3.3) at the 1939 World's Fair in New York, GM showed radio-controlled cars being moved by electromagnetic fields, which were generated by circuits embedded in the roadways themselves—later called intelligent highways. Nothing much came out of that effort; it remained only a vision.

*Figure 3.3: GM's 1939 Futurama exhibit about future cars and urban transit (Source: GM)*

## 3.1.3 RCA Labs and General Motors Get into Automated Driving Trials: the 1950s and 1960s[3]

In 1953, RCA Labs, in cooperation with GM, assembled a miniature car that was guided and controlled by wires embedded in a laboratory floor. This experiment struck a chord in the mind of a traffic engineer in the Nebraska Department of Roads by the name of Leland Hancock. Hancock decided to build a 400-foot-long experimental road strip in Lincoln, Nebraska. He buried experimental detector circuits, which looked like a series of lights, along the edge of the pavement. The detector circuits sent impulses to the car and determined the presence and velocity of any metallic vehicle on its surface. He developed his experiment in collaboration with GM, which equipped two standard models with special radio receivers and audible and visual warning devices that could simulate automatic steering, acceleration and brake control.

This capability was further demonstrated on June 5, 1960, at RCA Labs' headquarters in Princeton, New Jersey, where reporters were allowed to "drive" the cars. The system was expected to be commercialized by 1975, but it never was because there was no government funding for the road infrastructure.

Also in the 1950s and 1960s, GM showcased its Firebirds (shown in Figure 3.4), a series of experimental cars with an electronic guidance system that could move over an automatic highway with embedded metallic conductors.

*Figure 3.4: GM's Firebird with electronic brain (Source: GM)*

While these experiments were going on, the fertile minds of advertising executives were busy imagining a leisurely drive along a country road, a family playing a board game in an open, chauffeur-less limousine, as seen in the 1950 newspaper ad in Figure 3.5.

*Figure 3.5: An ad for a driverless car in 1950 (Source: GM Archives)*

### 3.1.4 Smart Highways Research: the 1960s and 1970s

During the 1960s, Ohio State University launched a "smart highway" project that built on the idea that driverless cars could be controlled by electronics embedded in highways.

Researchers predicted rather optimistically that such a roadway system could be in place in 15 years.

A few years later, four US states (Ohio, New York, California and Massachusetts) collaborated to propose an experimental electronic highway that would allow a special air-cushioned vehicle to drive (or, rather, hover) along it.

Around the same time, the UK Department of Transport was conducting similar research on its own with a Citroën DS that traveled on a road with a magnetic cable. The car, shown in Figure 3.6, traveled at a rather fast speed of 80 mph or 130km/h. The experiment demonstrated that automated driving was more consistent than human driving and could even handle bad weather. Cruise control devices were added later. The business case appeared strong—it promised a 50% increase in road capacity and a 40% reduction in accidents. Unfortunately, the funding was withdrawn in the mid-1970s.

*Figure 3.6: The Citroën DS (Source: UK Science Museum)*

### 3.1.5 The Stanford Lab Cart Project: 1964–71[4]

From 1964 to 1971, Stanford University worked on building a cart that could navigate itself around an uncharted "dumb" road (not a smart road embedded with electronics as in the previous experiments). The cart, shown in Figure 3.7, used early versions of autonomous-car artificial intelligence concepts for:
- Sensing its surroundings through machine (computer) vision.
- Processing information and making decisions.
- Making decisions and responding with corresponding cart movement.

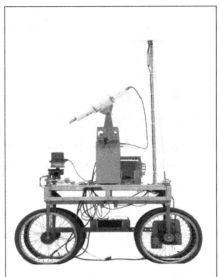

*Figure 3.7: The Stanford Cart, 1964–71 (Source: Stanford University, Mark Richards and ComputerHistory.org)*

The Stanford cart did wander into a nearby road and came back without causing an accident—an achievement worth noting and proving the underlying concept of self-driving in an academic setting.

### 3.1.6 CMU's Navlab Project: 1984[5]

Across the Atlantic in North America, CMU (Carnegie Mellon University) in Pittsburgh might deserve the distinction of spearheading, if not the theoretical research, then the application of technological knowledge and experience into prototypes of autonomous vehicles. Some call CMU the birthplace of AVs. In fact, CMU's efforts in the 1980s represented the second stage of the development of this major technological product wherein multidisciplinary teams (made up of mechanical and electronics engineers, robotics experts, AI neural network experts and systems-integration experts) cooperated to create a solution that could be put into real-life trials. CMU's website lists following projects under the program, which it calls Navlab:

**1984**: Navlab sets a goal to apply computer vision, sensors and high-speed processors to create vehicles that drive themselves
**1986**: Humans or computers controlled NavLab1, a Chevy van; the top speed achieved was 20 mph
**1990**: Navlab 2, a U.S. Army Humvee, wrangled rough terrain at 6 mph and highways at 70 mph
**1995**: Navlab 5, a Pontiac Trans Sport, traveled from Pittsburgh to San Diego in the "No Hands Across America" tour; the demonstration ride involved 3,100 miles with 98.2% autonomous steering control, with brakes and throttle human-controlled
**2000**: Navlab 11, a Jeep, was equipped with Virtual Valet - early version of autopilot
**2005**: Sandstorm and H1ghlander placed second and third in the DARPA Grand Challenge

**2007**: CMU's Boss vehicle won the DARPA Grand Urban Challenge along a 55-mile course

**2014**: CMU's latest self-driving vehicle is a Cadillac SRX that avoids pedestrians and cyclists, takes ramps and merges, and recognizes and obeys traffic lights

### 3.1.7 Europe's Self-Driving Efforts: 1987–1995[6]

Many people in the United States (including me, during my initial investigations of the subject) don't realize that Daimler Benz achieved a lot of success in semi-autonomous vehicles in the 1980s. Judging by the coverage in the press, the impression among North American audiences is that the autonomous car revolution began with Tesla and Google. In reality, German luxury car manufacturer Daimler Benz produced a prototype robotic van in 1986 with the help of Professor Ernst Dickmanns and his team at Bundeswehr University in Munich (see Figure 3.8). It could accelerate, brake, steer and make the trip from point A to point B by itself, without getting into an accident, on a real European autobahn. Professor Ernst Dickmanns is a pioneer in autonomous car research.

The largest collaborative industry/academia research project in the world, by far, for autonomous cars was called Eureka PROMETHEUS (Program for European Traffic with Highest Efficiency and Unprecedented Safety), which had funding of over 750 million euros. The program started in 1986 and several European manufacturers (led by Daimler), electronic suppliers, universities and engineering institutes were involved in this collaborative effort.

As a result of the project, Daimler produced several prototypes of semi-autonomous cars (level 2 by current SAE standards), culminating in a re-engineered W140 S-Class automobile that was able to drive almost by itself from Munich to Copenhagen—a distance of 1,678 km (1,043 miles)—in 1995. It reached speed of 175 kmph on the autobahn with a mean time between human interventions of 5.6 miles (9.6 km); this translates to 95% autonomous driving. Car's driving pedals (steering, throttle and brakes) were controlled by a computer after analyzing image data captured by four different cameras. The car also performed lane-changing function when it wanted to overtake other cars in its path.

A lot of this innovation in Europe came from Professor Ernst Dickmanns and his team at the University of Bundeswehr. Dickmanns outfitted a five-ton Mercedes truck with a bank of cameras and 60 computer modules to take pictures in search of obstructions (pedestrians, construction equipment, etc.) on the road, both in front of and behind the car. This was a very interesting application of the science of computer vision, capturing not just still images but dynamic ones. In a sense, he gave eyes to the car. The VaMoRs project allowed Dickmanns to later build appropriately configured semi-autonomous Mercedes cars that could travel on Germany's autobahns at high speeds.

*Figure 3.8: Professor Dickmanns's VaMoRs Mercedes van, 1986 (Source: Mercedes)*

### 3.1.8 The DARPA-Funded ALV Project in the United States: 1980s

During the 1980s, DARPA (the Defense Armed Research Project Agency of United States) funded the Autonomous Land Vehicle or ALV project using the latest AV technology coming out of universities, research institutes and companies (including CMU, the University of Maryland, the Environmental Research Institute of Michigan, Martin Marietta and SRI International). The ALV project was able to demonstrate vehicles equipped with LiDAR, computer vision and robotic control software travel at a speed of 19 mph.

### 3.1.9 The US Government's Formation of NAHSC and Demo97: 1991[7]

In 1991, the US federal government asked the US Department of Transportation (DOT) and the Federal Highway Administration to demonstrate an automated vehicle and highway system with an initial funding of $90 million. GM, Delco, Caltrans, Parsons Brinckerhoff, Bechtel, UC Berkeley, CMU and Lockheed Martin were major players in this consortium – National Automated Highway System Consortium (NAHSC). Their engineering and research work resulted in Demo97 in which they demonstrated 20 automated vehicles (cars, buses and trucks) operating in segregated and mixed traffic. This did not lead to commercialization, however, as the US DOT cut the funding in the late 1990s.

### 3.1.10 Daimler's Semi-Autonomous Vehicles Demonstration: 1994

In 1994, Daimler demonstrated twin robot vehicles, called VaMP and Vita-2, on a three-lane highway in Paris, achieving speeds of 81 mph (130 kmph) in heavy traffic.

### 3.1.11 Italian University of Parma's ARGO Project: 1996

Professor Alberto Broggi of the University of Parma in Italy launched the ARGO project, where the key emphasis was to have a modified Lancia Thema car follow painted lane marks on a roadway. The project ended with a journey of 1,200 miles on an Italian highway. The car operated autonomously for 94% of the time.

## 3.2 Autonomous Vehicles Start Taking Shape in the Twenty-First Century

Much research and experimentation for driverless vehicles was conducted in the twentieth century, but it was only in the twenty-first century that researchers started using AI extensively to automate driving.

### 3.2.1 The First DARPA Grand Challenge: 2004[8] *(Source of DARPA Challenge Content – DARPA website)*

"At the break of dawn on March 13, 2004, 15 vehicles left a starting gate in the desert outside of Barstow, Calif., to make AV history under the DARPA Grand Challenge project umbrella, a first-of-its-kind race to foster the development of self-driving ground vehicles. The immediate goal was to autonomously navigate a 142-mile dessert course that ran across from Barstow, California to Primm, Nevada. The longer-term aim was to accelerate development of the technological foundations for autonomous vehicles that could ultimately substitute for men and women in hazardous military operations, such as supply convoys.

The DARPA Grand Challenge was designed to reach beyond the traditional defense research community base and tap into the ingenuity of the wider research community in USA. It was DARPA's first major attempt to use a prize-based competition to attract novel performers and ideas - thus encourage collaboration across diverse fields. DARPA offered the first team to pass a series of qualification tests and then complete the course in less than the prescribed ten-hour time limit a $1 million cash prize.

The technological hurdles and rugged desert course proved to be too much for the teams' first attempt. None finished the course—the top-scoring vehicle traveled only 7.5 miles—and the prize went unclaimed. The competition wasn't a loss, however. It offered a promising glimpse of what was possible.

*Figure 3.9: Vehicles in the DARPA challenge (Source: DARPA website)*

"That first competition created a community of innovators, engineers, students, programmers, off-road racers, backyard mechanics, inventors and dreamers who came together to make history by trying to solve a tough technical problem," said Lt. Col. Scott Wadle, DARPA's liaison to the US Marine Corps. "The fresh thinking they brought was the spark that has triggered major advances in the development of autonomous robotic ground vehicle technology in the years since."

### 3.2.2 Second DARPA Grand Challenge (2005)

Just one day after the first challenge ended, DARPA announced it would hold a second Grand Challenge in the fall of 2005, 18 months after the first. This time, after analyzing lessons learned, five vehicles out of the 195 teams that entered successfully completed a 132-mile course in southern Nevada. Stanford University's entry, "Stanley" finished first with a time of 6 hours and 53 minutes and won the $2 million prize.

### 3.2.3 Third DARPA Grand Challenge (2007)

To further raise the bar, DARPA conducted a third competition, the Urban Challenge, in 2007 that featured driverless vehicles navigating a complex course in a staged city environment in Victorville, California., negotiating other moving traffic and obstacles while obeying traffic regulations. Six teams out of 11 successfully completed the course. The "Tartan Racing" team, led by Carnegie Mellon University, placed first in points awarded based on time to complete and ability to follow California driving rules and won the $2 million prize.

### 3.2.4 Italy's VisLab Intercontinental Autonomous Challenge: July 2010

Professor Broggi in Italy mounted 9,900-mile intercontinental challenge through Europe to China via Russia using autonomous vehicles. Four electric vans drove driverless for 100 days, although there were researchers in the vans who had to take control several times going through Moscow's traffic jams and toll booths. The exercise provided a lot of useful insight into the autonomous driving challenge.

## 3.3 More R&D Efforts and Test Runs by Auto OEMs: 2010 Onward

By the end of 2009, much had been researched, prototyped and tested in semi-autonomous car technology—both in the lab and on the road. Major auto OEMs like Daimler Mercedes and GM, academic R&D organizations and a few independent industrial research organizations had proven that component technologies required for autonomous cars do work. The missing element was software integration based on AI. The next phase of these efforts began in 2009 and is still going on. Let's discuss some of the more noteworthy efforts.

### 3.3.1 Audi

Audi built a TTS car that climbed driverless to the top of Colorado's 14,000-foot Pikes Peak in 27 minutes. A race-car driver could do it in 17 minutes but speed was not the most important factor—"no human driver" was the main achievement in this experiment.

### 3.3.2 VisLab: 2013

VisLab, a research company associated with the University of Parma, Italy, conducted another pioneering test in 2013. A specially-configured semi-autonomous vehicle called BRAiVE went around speed bumps, tight roundabouts, traffic signals, pedestrian crossings and other common road hazards on rural roads. It took the industry closer to its goal.

### 3.3.3 Volvo's Drive Me User Testing Based on Electric Plus Hybrid Power Train: 2017

Volvo, as a major European manufacturer, is well-known for building safe cars. Volvo has also made a commitment toward going green by announcing all of its **new** cars will be designed with an electric power train starting in 2019. This implies that Volvo's autonomous cars will be mostly battery powered. In April 2016, Volvo announced two separate plans to get feedback from real users on its autonomous technology. In 2017, the company announced that it would lease 100 XC90 plug-in hybrid cars with Drive Me technology (shown in Figure 3.11) to selected customers in London and another 100 electric cars to selected customers in China with the same Drive Me capability. In December 2017, after a partial rollout, Volvo scaled back on this effort.

### 3.3.4 Google, Tesla and Uber Enter the Autonomous Car Race

Established auto OEMs have been watching the AV scene and carrying on incremental R&D for the past ten years at a pace and risk-averse ratio that they might describe as pragmatic but that technology challengers like Google, Tesla and Uber might say is too conservative. It might also be misaligned with the reality of an imminent backdoor entry by Silicon Valley players into their empire. Google and Tesla have certainly shaken things up in the autonomous vehicle space since 2010. Uber, another upstart from the Silicon Valley has been operating a very popular ride-hailing service for several years since 2009. It set up collaborative and independent R&D activities in autonomous vehicle technology with an objective to offer future ride-hailing services without a human driver.

Google started its self-driving car project in stealth mode in 2009 with a modified Toyota Prius, but made it public in 2010. The project was the brainchild of company founder Sergey Brin. Google set out to drive the car fully autonomously over ten uninterrupted 100-mile routes. Within a period of a few months, Google's self-driving car (a Toyota Prius with Google technology) had succeeded in driving much farther. In 2012, Google outfitted a Toyota RX450 with the necessary autonomous gear and accumulated 300,000 miles worth of driving experience. While the company had some of its employees try the technology on highways, it also began gaining experience in more difficult road conditions—driving on busy streets and overcoming obstructions like road work and cyclists. In 2015, Google launched its famous Firefly car. To capture some headlines, it asked a blind man named Steve Mahon to try it out (in the company of a test driver).

*Figure 3.10: Google Lexus RX 450 trials (Source: Waymo)*

In 2016, Google carved out its self-driving car project into a separate company called Waymo. According to its website, Waymo had racked up three million real miles of self-driving and over a billion miles in simulation by the end of 2017. This translates to 25,000 miles per week. This might sound like a lot but after Uber's fatal crash in Arizona in March 2018, some experts are now predicting that a billion real driving miles by the industry might be a more appropriate target for trials before we can say that we have a well-tested AV that we can unleash with confidence to the public.

*Figure 3.11: Google's Firefly self-driving pod, born 2015 retired 2017 (Source: Waymo)*

### 3.3.5 The Waymo Self-Riding Program

In 2017, Waymo started a self-riding program in Phoenix, Arizona. The intent is to give Waymo's self-driving cars to pre-selected participants from the public who, in return, are expected to provide input to Waymo—what they like, what they don't, etc. A similar program was launched by Daimler Mercedes who invited auto journalists in Canada to ride their 2016 E320 series semi-autonomous cars and provide feedback.

### Google Field Trial Experience in 2015-2016 (Source wired.com)

The California DOT reported the following results for Waymo's self-driving car field trials during 2015–2016. Waymo's cars drove 636,000 miles with 124 disengagements (when autopilot software alerted the human driver to take over. This translates into roughly 500 miles per disengagement. It might be mentioned that most of this driving took place in Mountain View area in California that has relatively light traffic. Nonetheless it shows how far autonomous driving had progressed in 2016.

### How Google Outfitted Its Autonomous Car

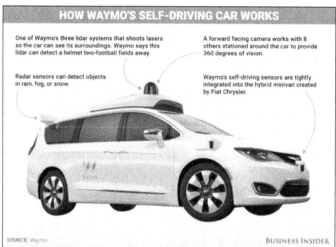

*Figure 3.12: How Google's self-driving car works (Source: Waymo and Business Insider)*

Google's technical head for the project has explained that the "heart of our system" is LiDAR—a combination of laser and radar. This is mounted on the roof of the car. Google uses LiDAR from Velodyne that sends out 64 beams of light to get information about the terrain. Information generated by laser pulses is used to develop 3D maps. It combines this real-time information from LiDAR and other sensors including cameras with high resolution maps of the road to generate current model of the terrain that route generation software uses to drive the car while avoiding obstacles and obeying traffic rules. The street price of the equipment on Google's self-driving car is very high: US$150,000 in total, with the LiDAR system itself at $75,000. The expectation is that as the cars are deployed in higher volumes, the LiDAR prices will come down drastically, to around $1,000.

There will be more information on this topic in Chapter 5.

### 3.3.6 Tesla Refocuses its Strategy, Making its Electric Car an Autonomous Car

Tesla has put a lot of pressure on existing OEMs to change their strategy to emphasize two important aspects of the future car: that the future AV Plus car should use electricity as a power source, and that self-driving hardware should be a standard feature in future cars (once the industry reaches a consensus on the minimum configuration). Once the hardware is in place, Tesla says, we can keep improving the self-driving capability through software updates.

Tesla's implementation of real-time updating of its auto-pilot software is ground-breaking. It is, in fact, a validation of my own prediction of such a concept as early as 1997 in my presentation on future wireless applications at the PacRim Comdex Computer Conference. I was, in that presentation, speculating on the potential use of wireless communication for enterprise applications, including the potential for auto companies to fix computer-related vehicle problems through an OBD (on board diagnostics) interface. I knew that the OBD computer module in cars was picking up all kinds of diagnostic information such as "out of sync" timing-belt engine-firing sequences. I understood that the timing belt in a car could become defective if it got out of sync, and I wondered whether this could be fixed by sending an informational message to the driver to attend to its immediate repair. I also speculated whether minor problems could actually be fixed by sending a "patch" command to the car's computers. I never thought at the time that one day an entire "autopilot" software could be loaded onto an AV computer. I wish I had patented the idea of AV software updating!

Elon Musk, the founder of Tesla, has certainly challenged auto industry leaders to change the course of the industry. Removing the world's dependence on the internal combustion engine that has been polluting the atmosphere for decades and replacing it with electric batteries has been a mantra for Musk. He captured the attention of the press, consumers and governments alike by conceiving, designing, manufacturing and delivering an all-electric car (the Tesla S-class) with a driving range that others couldn't fathom. Musk is a charismatic technology innovator who is willing to take huge risks and change the course of the industry. He's involved with electric cars, solar power farms, the LA tunnel plan and SpaceX ventures. Founded in 2003, Tesla is a unique, vertically-integrated auto manufacturing company that assembles its premium Model S (selling in the $80,000–$140,000 range), SUV Model X ($79,000) and the affordable Model 3 ($35,000 before subsidies) cars, which all use an all-electric power train, in its own plant in Fremont, California. Unlike other OEMs, Tesla designs and manufactures most of the components that go into its cars. Tesla Model S delivery began in 2012 and Model 3 deliveries began in July 2017. While Model 3 delivery is still sputtering, it did set a bar for the industry: an all-electric car with autonomous features built in.

Tesla's architecture is unique and a departure from the rest of the industry. Tesla was the first car company in history to have the ability to wirelessly download the control software that is the brain behind its cars' operation. Tesla cars use many of the advanced ADAS features available like automatic parking, lane control, autonomous steering, braking and speed adjustment.

Tesla announced in October 2014 that it would deliver its first version of Autopilot—a piece of software that would give it semi-autonomous capability—in 2015. The first version of Autopilot was delivered in October 2015 and several enhancements have been delivered since then. Note that Tesla's current Autopilot software is suitable for limited-access highway driving only. It can't detect obstructions like pedestrians and cyclists, so it's not suitable for urban driving. Essentially, as of October 2017, Tesla provided SAE level 2.5 autonomous capabilities on the highway. In Tesla's words, Autopilot just assists the human driver who must always be in control.

As far as full autonomy is concerned, Elon Musk has made some "off-the-cuff" announcements in the press and in a TED Talk suggesting that the company plans to deliver level-5 autonomous capabilities in its cars by 2019. Tesla is also planning a coast-to-coast test drive from a Fremont, California, parking lot to a New York parking lot in 2018 in which the driver will never touch the wheel. (Of course, entrepreneurs like Musk, however charismatic and capable of achieving impossible feats, are often overly optimistic. Elon Musk delivers on his vision but not necessarily on time.

**Tesla's Uniqueness as an Industry Innovator**
Tesla has introduced many significant differences and innovations into the auto marketplace. The company:

1. Was the first major outsider (not belonging to the established group of OEMs) that decided to build future electric cars with semi-autonomous features from start to finish.
2. Is more vertically integrated, controlling more elements of its supply chain, than established OEMs.
3. Set up charging stations on national highways to accelerate the time table for green-car adoption.
4. Builds only electric battery-powered cars, not hybrids.
5. Uses a higher level of robots and automation in its manufacturing plants than any other auto manufacturer.
6. Was the first company to put into practice "over the air" downloading of its software, using this ability to update the Autopilot software that operates its car in semi-autonomous mode.
7. Currently relies heavily on computer vision only for building a virtual picture of surroundings, unlike others who use multiple sensors including LiDAR.
8. Replacing the traditional dashboard with a unified electronic user interface integrated with smartphone on a large 17" touch screen controlling almost everything in the car.

### 3.3.7 Computers Invade the Automobile, Making the Driving Task Easier Since 1945

Auto industry innovators have been working to automate the tedious task of driving for a long time so that more and more people can enjoy it. This was primarily made possible through extensive use of computer technology; a modern car contains more than 60 computer systems (see Figure 3.16).

The first major automation in cars was the introduction of **automatic transmission**, which removed the need to change gears manually. (Of course the stick shift is still preferred by many car enthusiasts.) GM developed the first automatic transmission using hydraulic fluid in the 1930s, and it introduced the "Hydramatic" transmission in 1940. The 1948 Oldsmobile was the first model to use a true automatic transmission.

Another significant automation was the invention of **cruise control** in the 1960s. This enhancement of the driving experience was (and is) most useful in highway driving over long stretches of relatively empty road. I remember how my foot used to get tired from staying in the same position for a long time; I was very happy when I bought a car with cruise control. Modern cruise control (which was called *speedostat* by early auto engineers) was invented in 1948 by a mechanical engineer named Ralph Teetor. Teetor's idea was born out of his frustration when riding in a car driven by his lawyer, who kept speeding up and slowing down as he talked. The first cars to boast the new technology were the 1958 models of Chrysler's Imperial, New Yorker and Windsor. By 1960, cruise control was a standard feature on all Cadillacs. The system worked by calculating ground speed based on driveshaft rotations.

As embedded computers have gotten more and more sophisticated in real-time machine control applications, so have automotive applications of these computers. Today we are seeing a rapid acceleration of these types of innovations for high-end cars. Regular cruise control has now been replaced by **ACC (adaptive cruise control)**, which is becoming quite common in today's semi-autonomous luxury cars. These cars use either a single sensor or combination of sensors (radar and camera) to allow the vehicle to maintain a constant distance from the car in front of it, slowing down when closing in on the vehicle ahead and accelerating again to the preset speed when traffic allows. Some systems also feature forward-collision warning systems, which warn the driver if a vehicle in front gets too close (within the preset headway or braking distance).

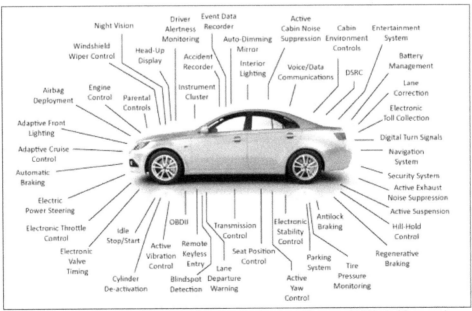

*Figure 3.13: There are 60+ computers in a modern car (Source: QNXAuto.blog.com)*

**Automatic braking** is a safety technology that activates the vehicle's brake system, to some degree, when necessary. The level of automation varies from pre-charging the brakes, to slowing the vehicle to lessen damage, to taking over completely to try to stop the vehicle before a collision occurs. The Insurance Institute for Highway Safety says that automatic braking or brake assist is an integral component of crash-avoidance technologies, which include systems designed to prevent front crashes and backovers as well as cross-traffic alert systems. Each auto maker will have a different name for this technology, but the bottom line is that automatic braking is meant to minimize accidents.

**Active crash avoidance** technologies use sensors, cameras, radar, and LiDAR to detect vehicles. Most systems issue a visual or audible warning when detecting an impending vehicle collision. Some also provide brake assistance, pre-charging the brakes to provide more power when the driver hits the brake pedal. If the driver fails to brake, cars with active crash avoidance feature are designed to brake on their own. Some brake systems immediately apply the brakes without issuing a warning to the driver. But not all auto-braking technologies are meant to completely prevent a crash; the primary goal of some systems is simply to slow the vehicle, to make the impending impact less severe.

Many of the safety-enhancing capabilities described here use sensors that detect potential obstructions or other vehicles (in front and behind) as they come too close to the car, and then feed that information to the car's embedded computers, which make decisions based on the received information. Those decisions are sent to devices that control the level of throttle for acceleration or deceleration, apply brakes and control steering.

### 3.3.8 Professor Hinton's Research at the University of Toronto Gives New Hope for Fully Autonomous Vehicles

In my introduction to the chapter, I mentioned that Professor Hinton's research in AI at the University of Toronto during 2010-2012 allowed scientists to recognize pedestrians and other obstructions on the road using computer visualization technology. This was an important milestone in our journey towards autonomous vehicles. It allowed other AI researchers to carry his work forward. Subsequently Professor Hinton was hired by Google Brain Lab in Toronto where he continues to do pioneering work in this area, along with his academic work at the university.

## 3.4 Reviewing Our Progress over the Past Century

Our historical journey toward autonomous capabilities over the past century is summarized in Figure 3.17.

It took us a long time to get this far. Researchers and their compatriots in industry tried many approaches. Initially, they tried radio waves (as in the Houdina experiments) as a means of controlling cars remotely; that approach had its shortcomings and was soon dropped. Then, they tried embedding electronics in roadways in an effort to make the roads and highways intelligent (GM, RCA Labs, Ohio State, the UK government, etc.).

Many jurisdictions in the United States tried to make a case for intelligent highways and even got a few politicians excited about it, despite the fact that the technology hadn't been proven and the benefits had not been quantified. This was an ambitious and expensive approach indeed—it would have involved huge expenditures. This idea, too, was dropped, in the 1960s. Then, in the mid-1960s and 1970s, academic labs like Stanford, CMU and Germany's Bundeswehr University reversed the concept of where the autonomous cars' intelligence should be: it should be not in the road but in the vehicle. What if electronics in the car, they wondered, could detect the outside environment and use software to determine how to operate the car? In this respect, Professor Dickmanns' work in Germany was ground-breaking. From the 1980s onward, Europe took the lead with its Eureka PROMETHEUS project and spent a lot of money and brain power to test concepts and technology components that would become the underpinnings of our current approach to autonomous cars.

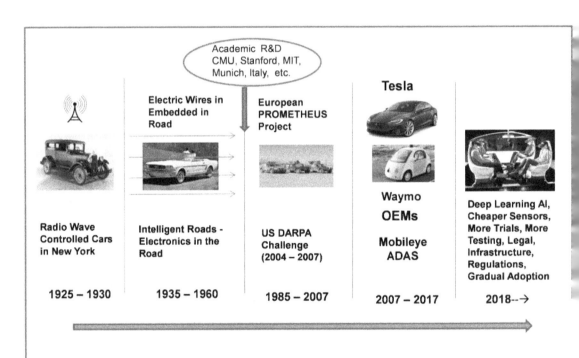

Figure 3.14: History of autonomous vehicles—How we got here

Research, engineering development, retrofitting cars with sensors, research competitions like DARPA and on-the-road trials in the 1980s, 1990s and early 2000s brought great achievements to the industry. It was clear by the end of 2010 that it was possible to produce significant level of autonomy if the deep-learning discipline of AI could be used to mimic a human driver accurately. Whether we'll achieve SAE level-5 autonomy in the next decade is a question that I'll debate further in Part 4 of this book. But it's quite evident from the journey so far that we are now on the right track. We can see some rays of light at the end of the tunnel, although we can't yet accurately measure that tunnel's length.

### 3.4.1 How Are We Doing Safety-Wise So Far?

While NHTSA in the United States is responsible for monitoring safety on highways, we do not have a comprehensive picture of the safety record of various trials going on. Accidents involving autonomous vehicles do make the news. Press coverage of the following accidents put us on alert indicating that we are not there yet.

Google has provided some information on its safety record with self-driving cars. In July 2015, Google pointed out that since the start of the project in 2009, its test vehicles involved in the self-driving project were involved in 14 minor accidents – with no casualties. Chris Urmson, Google's Head of the project pointed out that most of these accidents were caused by humans driving other cars. Interestingly, 11 out of 14 were rear end accidents. Urmson said - "Our self-driving cars are being hit surprisingly often by other drivers who are distracted and not paying attention to the road." He further indicated that the test vehicles had driven more than two million miles over a period of six years (2009 – 2015).

The first known fatal accident involving a self-driving vehicle took place in Williston, Florida, on May 7, 2016, while a Tesla Model-S electric car was engaged in Autopilot mode. The driver of Tesla car died at the scene when his car hit an 18-wheeler tractor-trailer coming from the opposite direction made a left turn at an intersection on a non-controlled access highway. According to NHTSA, neither the driver nor the auto-pilot of Tesla Model-S saw the white side of the tractor trailer against a brightly-lit sky. The car did not apply brakes and went full speed under the trailer.

In March 2018, there were two fatal accidents involving self-driving cars. One involved an Uber autonomous vehicle (which had a human driver in the car for safety) where a 49-year-old pedestrian was killed in Tempe, Arizona. Preliminary indications were that the Autopilot in the Uber AV didn't notice the pedestrian and didn't slow down. The second accident involved a Tesla Model X in Mountain View, California. Tesla confirmed that Autopilot was engaged and that the human driver did not respond to the system's request for the driver to take control when an obstruction (a divided highway barrier) was in the car's path. The question of blame is complex in these situations—do we blame the Autopilot, the human driver ... or perhaps both?

These accidents have highlighted the need for more thorough testing and have raised a red flag for the AV industry, suggesting that they're moving too fast and introducing self-driving technology without enough testing and certification. I'll discuss this topic of the maturity of AV technology in Chapters 12 and 14.

## 3.5 AV Development Continues—Where We Were at the End of 2018

It's clear that the industry had gone past the "concept and R&D" stage for semi-autonomous cars (SAE level 2) by 2015. Public trials of these AVs began on public roads (admittedly somewhat protected and controlled roads in some cases) in 2016–2017. These trials (which stopped in 2018 after the fatal accidents discussed in the previous section) were being conducted by at least 10 vendors: a combination of established OEMs, newcomers like Google and Tesla, ride-hailing operators like Uber and Lyft, and core AV component suppliers like Intel Mobileye, Nvidia and others. The results were generally positive—problems were found and being fixed, sensor hardware was becoming cheaper and AI software was being incrementally improved. The following results attest to this continuous progress:

- On the R&D side, many breakthroughs required to achieve the vision of fully autonomous cars had been achieved, except for the ability of AI to mimic a human driver in all respects and for all situations. Researchers at MIT, Stanford, CMU and European research labs like the University of Bundeswehr in Munich are expressing cautious optimism that they will be able to replace the human driver through deep-learning AI in due course.
- On the technology-trial front, major established OEMs (GM, Ford, VW, Mercedes, and BMW) have prototype AVs in commercial trials in the United States and/or Europe. Even though the current AVs are capable of SAE level 2 or, at best, SAE level 3, the majority of vendors do have a strategy (or at least a target) to unveil fully autonomous cars as early as 2019 (Tesla) and as late as 2029.
- On the silicon side, cheaper sensors are being developed. Intel/Mobileye and Nvidia are developing high-performance hardware for handling the massive amounts of data that these cars will need to process in real time.
- There's no shortage of VCs and investors ready to provide the capital to make this dream come true.
- On the legislative side, the US Congress and Senate are supportive of the AV effort. There is bipartisan support in the United States and elsewhere. However, in my view, blanket support without adequate regulations, including suitable penalties, is a recipe for disaster.
- A number of independent test initiatives/facilities are springing up throughout the developed world, such as the American Center of Mobility in Michigan.
- Many business partnerships between established OEMs and ride-hailing and car-rental agencies are being formed in an effort to make TaaS a reality. Operational mechanisms and pricing models are still uncertain.
- Some cities have launched studies to understand the impact of AVs on the road infrastructure. Funding and time are, of course, major obstacles.

Some of this sounds promising, some daunting. The question is: is it enough to unleash such a "life and death" technology without enough safeguards during the next decade?

**Summary**
*We've come a long way from our initial attempts in 1925 to achieve self-driving capability. We made two major detours along the way – switching from radio-controlled cars to embedding wires in the roads. Finally we seem to be on the right track—putting sensors, computers and software into the cars themselves. We can see the destination but we can also see lots of hurdles, potholes and puddles in our way.*

**Citations for External References**

[3] Smart Highways and Roads by RCA & GM –
https://www.engineering.com/DesignerEdge/DesignerEdgeArticles/ArticleID/12665/The-Road-to-Driverless-Cars-1925--2025.aspx

[4] Stanford University Cart Project – https://web.stanford.edu/~learnest/les/cart.html

[5] CMU Navlab Initiatives – https://www.cs.cmu.edu/afs/cs/project/alv/www/

[6] Europe's Eureka Prometheus Project – https://en.wikipedia.org/wiki/Eureka_Prometheus_Project

[7] NAHSC and Demo97 – https://www.fhwa.dot.gov/publications/publicroads/97july/demo97.cfm

[8] DARPA AV Challenge – https://www.darpa.mil/news-events/2014-03-13

# Chapter 4

# The Rationale for AV Plus

> *I'm not sure what the true motivations were when Houdina engineers built their first self-driving prototype in New York in 1925, but clearly self-driving cars have always generated a lot of excitement. We do know that training human beings to drive cars without causing accidents and deaths has always been a challenging task. So early auto engineers believed that taking the driver out of the equation could be not only exciting but also safer—surely a recipe for riches! As time went on, new motivations emerged: improving congestion in cities, reducing commute time, improving people productivity and enabling people with disabilities move from one place to another.*

There is a law of probability that applies to the movement of cars on shared roads. When two or more objects are moving in the same plane and in the same vicinity without following prescribed rules and each of these objects thinks it has the right of way, they are likely to collide once in a while.

The automobile was invented in 1885 by Karl Benz in Germany and motorized vehicles were introduced on public roads in the early 1890s. It wasn't long after that accidents and the resultant injuries and even deaths started happening. In the early days of the auto revolution, drivers were inexperienced, vehicle speeds were low, road infrastructure was poor, traffic lights did not exist, there were mixed vehicle types on the road and the rules of the road had not been established. Even today, the traffic scene is not so different from this in congested cities like New Delhi.

Since the early days of the automobile, automakers have been trying to improve the safety and drivability of their cars. Regulators and legislators in America and Europe have often called upon manufacturers to make the safety of drivers, passengers, pedestrians and other vehicles on the road a critical consideration in their designs.

According to *Detroit News Archives*, early days of the automobile (1900–1930) were a period of dangerous driving. The picture in Figure 4.1 shows a typical scene from those days. Initially there were very few cars. In 1909, there were only 200,000 motorcars in the United States; this increased to over a million cars in 1916. Drivers were inexperienced; there were no traffic lights and no traffic signs. Automobile drivers, tram drivers, horse-drawn carriage drivers and pedestrians had to look out for themselves. Naturally, there were lots of accidents. In fact, in 1917, according to the *Detroit News*, there were 65,000 cars in Detroit and 7,171 accidents, of which 168 were fatal.

Figure 4.1: A typical traffic scene from the early days of the automobile (Source: Detroit News archives)[9]

Soon after a period of unregulated traffic, Detroit led the way in the United States in putting order to the chaos by instituting road signs and road markings, along with traffic police to administer the rules. However, the problem grew worse as paved roads were built and driving speeds increased. Instead of fender benders in early days, it became more common for accidents to result in serious injury or death. Driver training and licensing of drivers helped but increasing number of vehicles on the road led to an increasing number of accidents. Safety, therefore, has always been a major issue for customers and auto manufacturers alike.

## 4.1 The Single Biggest Rationale: Reducing Accidents and Deaths

While cars have become safer over the years with seat belts, airbags and other safety features, deaths due to automobile accidents caused by human error continue to be the single most important justification for AVs. The National Safety Council (NSC) reported that 40,000 people died in the United States alone in motor vehicle crashes in 2016, a 6% rise from 2015 figures. The World Health Organization estimated that worldwide there were 1.25 million deaths due to road traffic accidents in 2013.

Preventing loss of life is, of course, the most emotional rationale for AVs. However, motor vehicle accidents may or may not involve death. Instead these may cause other types of injuries and economic losses. The NSC estimates that the *economic* loss due to motor vehicle–related accidents in 2016 was $432 billion, which includes losses due to death, wage and productivity losses, healthcare costs, property loss, employer costs and administrative expenses.

How do you reduce this human tragedy and economic loss? **Is it not a dichotomy that the developed world loses more people to its own technological inventions than it does in ongoing wars that the developed world is involved in?** Until recently, the automakers have put a big emphasis on making cars safer. Statistics show that the introduction of seat belts in the 1970s led to a substantial reduction in injuries. According to CDC and Insurance Institute of Highway Safety websites, seat belts reduce the risk of death by 45%, and cut the risk of serious injury by 50%. Automakers have recently begun focusing on innovations that assist the driver in a number of other ways, such as blind-spot assist, lane-keeping assist, adaptive cruise control and emergency braking. The idea of replacing the human driver who's responsible for the majority of traffic accidents with an AI-controlled autopilot is a quantum step forward. **Nonetheless, this is a major motivation for today's push toward fully autonomous cars.**

The typical human driver is a risky driver. It's just a fact that we're not fully attentive and focused on the road 100% of the time. We're easily distracted by many things—talking to other passengers, drinking our coffees, switching radio stations, fiddling with climate controls, yelling at the kids or talking on the phone with the boss. Our minds are always multi-tasking, and we operate our cars almost without thinking. We may not always have our eyes on the road. We can't always see the traffic in all lanes or spot jay-walking pedestrians. Our reaction time can be too long. Some people, unfortunately, drive under the influence drugs and alcohol. It's an imperfect system no matter how you look at it.

On the other hand, an AI-powered autopilot can see everything, is fully attentive and is completely dedicated to the singular task of driving safely. Studies suggest that AVs could significantly reduce human deaths due to traffic accidents. A 2015 McKinsey article suggested that if all cars were replaced by AVs and human drivers were taken off the road, it could reduce car accidents by as much as 90%. Assuming that translated to a 90% reduction in deaths, widespread AV adoption could prevent 36,000 deaths (NSC figures) in the US, and one million deaths worldwide, each year. Setting emotions aside, accountants, lawyers and actuaries assign a dollar value to human lives and injuries; based on those numbers, 36,000 lives saved annually would translate to US$190 billion in savings in healthcare, liability claims and loss of income.

I believe that these estimates by McKinsey[10] are based on unrealistic assumptions and very optimistic outcomes where nothing goes wrong—an unlikely scenario. I haven't seen the assumptions behind these estimates, and I feel that they're are based on speculative analysis about an environment that has yet to play out. Some observers believe that things could get somewhat chaotic in the initial transitional period of mixed environment where human drivers do not know how to deal with driverless vehicles on the road and thus cause more accidents. Nevertheless, this would gradually settle down as human drivers and AVs learned how to co-exist, and I believe there *is* a potential for a significant reduction in deaths due to accidents when autonomous vehicles become the vehicles of choice by the majority of consumers.

## 4.2 Driving Time Savings and Productivity Enhancements

McKinsey report cited earlier estimates that city passengers could save as much as 50 minutes per day if AVs became the default mode of transportation. This is the average time that a person spends in the car driving to and from work in urban areas. What would we do with this newfound spare time? Some professionals would be more productive, going through their email, sending messages and reviewing presentations while being chauffeured by the robo-driver. Service people could use the time for administrative tasks like preparing invoices. Many others would just use the time to enjoy the scenery, watch some TV, take a nap or watch the ball game. The bottom line is that passengers could make better use of the time they spend going from point A to point B, whether for work or leisure.

## 4.3 Transportation as a Service and Reduction in Cars

AV enthusiasts suggest that once cars become fully autonomous, ride-sharing services will become more affordable and will be available at a moment's notice with just a click or voice command to the smartphone. This will mean there will be less economic motivation to own our own car(s). Unlike the current Uber-like services that use human drivers, you could book an AV for an hour or two or longer, as needed. Unlike today's leases, you could lease an AV for a bundle of trips or a certain number of miles per month. In fact, similar transportation services are available today (such as Zipcar and car2go) on a fairly similar business model in larger cities. The difference in the future would be that more options would be available and the car could come to your house without a driver and on short notice, because there would be lots of TaaS vehicles available in the vicinity.

The industry estimates that labor makes up 50% of our total transportation-service costs. Take the cost of the taxi driver or the Uber/Lyft driver out of the equation, and the per-mile or per-kilometer cost will come down significantly. Once we have electric cars, fuel costs will go down as well. Making cars safer will drive down insurance and liability costs. So AV-based transportation-on-demand services could come at a lower price than the cost of private car ownership or current forms of ride-sharing services. While TaaS appears very attractive, there are cost model assumptions that need a detailed analysis. I shall discuss that in Chapter 8.

Certainly, the number of cars in a household will decrease—a household may own one primary car that is then supplemented by an additional TaaS subscription. Some studies predict a rather optimistic 50% reduction in cars (essentially all households going from two cars to one) in urban areas. Independent think tank RethinkX report author Tony Seba predicts that the current population of cars in the United States will drop from 247 million to 44 million. I think that level of reduction is highly unlikely, but the number of cars on the road will definitely go down by some margin as AVs become the default over the next two generations.

## 4.4 Faster Trip Times

As I've mentioned, future transportation models have the potential to reduce traffic, especially in congested cities. Today the average vehicle utilization is 4–6%, but future TaaS AVs could be driven 24/7 at an 80–90% utilization level. As we reduce the number of cars on the road and also reduce accidents by replacing human drivers with autopilot software, we'll have fewer traffic snarls, which means trip time from point A to point B will go down and become more predictable.

As we won't own our own cars that need to be parked somewhere, we'll further reduce congestion (it has been estimated that in downtown areas, a good percentage [up to 30%] of traffic is caused by cars cruising around to find parking spots). Even privately owned AVs can let us off at our destinations and then go park themselves in designated parking spots or in public parking spaces. And of course if we're using a TaaS service from a ride-sharing company, the car can drop us off and then go pick up another passenger. All of these scenarios lead to less congestion on the roads. The integration of AVs with the public transportation system could further reduce congestion.

Not only will our trips be shorter, but they'll also be more enjoyable since we'll no longer experience the stress of driving, say AV promoters. This contradicts recent studies by Texas Tech University that suggest that our stress levels may *not* decrease for private AV owners, because of the need to stay vigilant and be ready to take over in case of an emergency.

## 4.5 Lower Insurance Costs

Reducing the number of accidents and casualties will lead to lower insurance costs. I'll discuss this further in Chapter 10.

## 4.6 Less Pollution in the Air

One of the most significant rationales for electric AVs is that they are cleaner, more reliable and better for the health of our descendants and our planet. Bloomberg estimates that 54% of cars manufactured will be electric by 2040 and that 33% of all cars on the road in the developed and developing world will then be electric. Assuming an average car has a lifespan of 12 years, the majority of cars on the road will be electric sometimes after 2052 emitting very low amounts of fossil-fuel-based pollutants. Granted, electricity-generating plants may still be using fossil fuels and we'll need this electricity to charge our cars, but still carbon pollutants could be reduced by at least 50%.

Since there will be fewer cars, and those we do have will be greener electric cars running autonomously and thus requiring no cruising around for parking spots, our children and grandchildren will have cleaner air to breathe—a huge benefit indeed.

## 4.7 Reduced Parking-Space Requirements

Today, parking requirements are quite onerous. Municipal zoning codes dictate that there must be a certain number of parking spaces based on the number of employees or occupants of a building (which can be huge for office complexes, hotels, convention centers and the like). Parking lots cost a bundle to build and maintain, and they take up a lot of space. Even in detached homes of 2,000+ square feet, we often see significant space allocated for garages (as much as 350 square feet for two cars). If AVs become mainstream, we could need less than half as much parking space. In fact, McKinsey estimates that driverless cars will reduce the need for parking space in the United States by more than 5.7 billion square meters. This could make for a significant savings in real-estate costs.

## 4.8 Increased Mobility for Seniors and People with Disabilities

Driver-assist technologies in the interim and driverless AVs ultimately could be a boon for the elderly or for people with disabilities. When Google had a blind person ride in one of its self-driving cars in 2012, it was a great public relations exercise—and really drove the point home.

*Figure 4.2: A blind man using Google self-driving car (Source: Waymo)*

I can't overstress the social benefits of AVs to this segment of our population. AVs could simply make their lives more meaningful. I could relate a personal true story to attest to this benefit. Our family knew an 80-year-old single woman who drove until her doctor observed one day that she could be a hazard on the road. On her next visit, the doctor took the car keys from her and gave those keys to my wife who looked after her financial investments. The woman was greatly disturbed by this action. My wife and I were advised to take the car away from her. We sold the car on her behalf. However, since that day, her mobility was greatly reduced and she grew lonely since she was unable to visit her friends and relatives. She was admitted to an old age home, where she sadly died within a year. Such is the desire of seniors to have freedom of physical mobility. A fully autonomous vehicle might have enabled her to continue to get around, stay active and, who knows, maybe even live a while longer—or at least get a little more enjoyment out of her life.

## 4.9 Fewer Repairs with AVs

Today's ICE (internal combustion engine)-based cars have a lot of moving parts. Future AVs would be predominantly electric and have fewer of these parts. Future AVs will therefore be more reliable than the present generation of vehicles.

Tesla claims that its Model S requires less planned and unplanned maintenance than gasoline-powered cars, and offers an eight-year warranty with unlimited mileage. Tesla's argument is that its cars are simpler in construction and have fewer mechanical parts that are subject to failure. There are no oil changes and no emissions checks required. Of course, tires and brake pads will need replacement like in other cars, but even brake pads will last longer because electric cars make extensive use of regenerative brakes, in which the engine does much of the work to slow the car. Electric cars don't have automatic transmissions, valves, spark plugs, crankshafts, connecting rods, cylinders, camshafts, exhaust gas recirculation systems, belt/pulley systems, oil pumps or engine cooling fans. There are even fewer ICs (integrated circuits) n electric cars than in gasoline cars. Of course, both varieties of AVs (ICE and electric) will have additional sensors, lasers, cameras and supporting computer gadgetry to support their autonomous capabilities, which could increase the cost of each service call, as highly skilled technicians will be required.

I should point out that higher reliability has to be designed and built *into* the vehicle, which does increase the cost of the vehicle itself. If OEMs establish higher reliability targets for AVs, they *can* be achieved. But this will require that these targets be considered while critical component specifications are developed and procured from external suppliers. Testing will have to be absolutely scrupulous and independent of vendor influences. If OEMs want to cut costs, higher quality will not be achieved.

## 4.10 Driving Boredom Out of the Window

I'm a baby boomer. I grew up with a car as an important part of my life. A car gives me physical mobility. It gives me the freedom to go anywhere I like. I like to have the convenience of a car, even in Toronto where we have a decent subway and public transportation system. (I'll admit that a car even provides an outlet for me to demonstrate my professional success by driving the type of car I can afford.) Besides allowing me to do necessary things like driving to and from work, it allows me to visit friends and take my children on long drives on highways (often using cruise control).

When my wife and I were bringing up our children, we often drove to New York, where my brother lived, for family get-togethers. Taking my daughters to the skating rink and on skiing trips gave me a great opportunity to bond with them. We enjoyed every one of our car trips, despite my millennial children's constant refrain : "Are we there yet?"

Driving was freedom and fun for us baby boomers. We would sometimes drive along the lake for no other reason than to watch and enjoy the lake. I wasn't a sports-car fan but I had a close friend who bought a Corvette in 1980 and I enjoyed a few rides in the countryside with him, with the roof down.

Baby boomers like me may find the ADAS features of semi-autonomous cars attractive. (That is why I invested in my new 2016 Mercedes C300 with "Intelligent Drive"... I have to admit, it's pretty great.) But many in my generation won't want to give up control entirely. My children, on the other hand, have much less interest in owning a car than my wife and I do, our grandchildren even less. Driving is boring for them. To them, the concept of driving-as-fun is passé. A car is only a necessity to them, a means of getting from point A to point B. Driving keeps them from other important pursuits like texting, emailing, Tweeting, WhatsApp-ing and Facebooking. AVs seem like just the right mode of transportation for these next generations.

Interestingly, the visionaries at Waymo and Tesla are millennials and the "wheelers and dealers" at established auto OEMs tend to be baby boomers. Who is driving whom—we can only speculate.

## 4.11 Reduced Car Ownership

Under our current model, owning a car for transportation makes good economic sense for most individuals and families. There are a number of reasons for this.

Taxi services are expensive. Uber-type services are only marginally cheaper. Public transport is cheaper but very inconvenient. Using Google Maps' estimates, I often find that a trip by public transportation is 3–4 times as long as going by car. Taxi services are expensive because of the costs of labor, fuel, insurance and vehicle maintenance.

If AVs don't need a human driver and have lower operating costs, lower insurance rates and lower maintenance costs, the old business model falls apart. The new business model, using AVs for on-demand transportation services, will make it economical to give up the ownership of a private car in some cases.

The economics and convenience behind the TaaS model are attractive—some business people see this as a huge disruption to our current transportation methods, both public and private. RethinkX's report entitled "Rethinking Transportation" to this effect has been much cited in the media; while it's an interesting read, I feel that it presents a highly optimistic outlook and a flawed business model. I believe it'll take much longer for TaaS using AVs to become the norm, and that many baby boomers and millennials alike will opt instead for private AVs when they become available. Affluent consumers may use a combination of private AV and TaaS for the next two or three decades, but it will be a very long time before we are willing to give up car ownership entirely, if we ever do.

## 4.12 More Efficient Transportation on Private Campuses[11]

We recognize that fully autonomous cars will require many more trials and testing for another decade or longer. Allowing fully autonomous cars in busy streets in congested cities has still many challenges to meet and solve before we allow fully autonomous cars onto busy streets in congested cities. However, what might be easier is allowing driverless cars, vans, fleet trucks and specialized motorized equipment (like construction or mining equipment) on private and protected campuses. On these sites, traffic is controlled and routes are well-marked. We already have trains operating on exclusive tracks in airports. Such use will be among the first applications of AVs.

Experiments with fleets of AVs are showing encouraging results. The City of Michigan started using Navya (a French AV fleet start-up) to move students and staff on minibuses around the Michigan University campus at the end of 2017. Early indications are the experiment is successful. University students (Z generation) are expected to be early adopters of AVs.

## 4.13 Saving Fuel and Low-Cost Shipping

According to MIT news[12], "researchers believe that autonomous vehicles may save fuel by as much as 15–20% by trailing each other in large platoons. Like birds and fighter jets flying in formation, or bikers and race car drivers drafting in packs, vehicles experience less aerodynamic drag when they drive close together."

As a longer-term application of automation in the transportation and package delivery space, the industry is interested in partially or fully removing the driver from supply chain applications. Through a hybrid van/drone solution, pizza delivery can become much cheaper, for example. Aggressive upstarts like Uber and Lyft are looking at this type of solution.

## 4.14 Advancing the State of the Art in AI and Deep Learning

One of the biggest indirect benefits that will occur is the advancement of AI applications in industrial (first) and consumer (second) applications. Many experienced technology watchers believe that fully autonomous AVs represent a huge challenge to AI in general and deep learning in particular, and they liken it to America's Apollo moon landing in 1969. Developing a fully autonomous vehicle that can mimic a human driver not only in seeing its surroundings but also in anticipating the intentions of human drivers and pedestrians in the vicinity is as difficult a challenge to the AI community as was the moon-shot challenge President Kennedy posed to the American scientific community in his famous 1961 speech. Just as the real moon shot advanced the state of the art in many industrial sectors and areas of scientific research, AVs have the potential of advancing AI research and applications over the next two decades.

## 4.15 Public Safety and Smart City Benefits

If the municipal infrastructure is part of our connected AV-transportation world, we could see huge benefits to public safety. Information about accidents could reach police, fire and medical agencies instantly, and responders could be immediately dispatched. Of course, cities would have to invest in upgrading their infrastructure. The US federal government has given $40 million to the city of Columbus and Ohio State University (supplemented by another $10 million from a venture-capital fund) to do research in a "smart city" area vis-à-vis AVs. It may be pointed out that fire and EMS services are integral part of smart city concept. For more information, go here.

## 4.16 Safer Use of Smartphones

The widespread use of smartphone use while driving has led public safety agencies across North America and Europe to come down hard on offenders during the past three years. Recent studies show that distracted driving is the single most common cause of accidents. Many states and provinces have outlawed distracted driving and are imposing both fines and demerit points for it. The use of hands-free Bluetooth connections in cars has helped to make phone calls safer, but texting and emailing continues to be a major hazard in current implementations.

Yet smartphone users have a strong desire to make calls, text or use other apps while driving. Seeing pedestrians cross the road with their eyes glued to their phones is now a common phenomenon. It's becoming quite clear that people want to use their smartphones 24x7, including the time in the car. Users are, therefore, starting to ask for Android Auto and Apple CarPlay to expand their in-car application capabilities. It would be better if it is integrated with the built in screen that comes with AVs. In fact, AV integration with smartphones will be almost essential in any future scenario, especially with SAE level 3 or 4 AVs. For safety reasons, initially, only a subset of smartphone apps should be supported with AVs.

## 4.17 Negative Aspects of AV Plus Disruption—What Price Will We Pay?

Is the outlook for AV revolution all positive? Unfortunately, there is no free lunch. The rewards of the innovation, fun and convenience of AVs come with a big price tag for society.

Let me re-emphasize the fact that apart from reducing accidents and congestion in cities, the major motivation for AVs is the removal of the human driver. This will inherently mean a loss of jobs in many sectors, in some cases in the millions. As this disruption plays out completely over the next 50 years, we will likely see significant job losses among taxi drivers (there are 240,000 today in the United States alone), long-haul truckers (there are 1.6 million in the United States today), and delivery-truck drivers (800,000 are employed by services like UPS and FedEx in the United States).

Since the future AVs will be complex machines with steel/metal/plastic body components, batteries, motors, embedded computers and sophisticated AI-centric software, servicing them will be a complicated task. Customers will tend to rely on established OEM dealers as well as new entrants, rather than local service shops. Some maintenance will even be done wirelessly and remotely. The current independent service technicians will find it very hard, in the short to medium term, to compete, unless they retrain and hire a new breed of technicians. Also, fewer technicians will be required because of the superior reliability of future cars. According to the US Department of Labor, there were 739,000 auto technicians employed in 2014. These auto technicians should upgrade their skill set.

As our future cars will be constantly connected and tracked, there will also be privacy implications too. Many won't be comfortable with this Orwellian future; this will need to be carefully considered. And finally, the cybersecurity of connected cars will be a major problem that will require careful design considerations. As of now, the industry has not addressed this issue seriously.

## Summary

*There is a strong justification for AVs as a long-term solution to vehicle accidents and the associated loss of human life. Many other benefits exist as well, although they're less significant. The key question is how we'll manage the transition from a human-driver regime to one characterized by AV Plus autopilots. I'll explore this subject from various angles throughout the book.*

---

### Citations for External References

[9] Detroit News Archives *http://www.detroitnews.com* Bill Loomis Reporter
http://www.detroitnews.com/story/news/local/michigan-history/2015/04/26/auto-traffic-history-detroit/26312107/

[10] McKinsey Report on Benefits of AVs – https://www.mckinsey.com/industries/automotive-and-assembly/our-insights/ten-ways-autonomous-driving-could-redefine-the-automotive-world

[11] University of Michigan and Navya – https://news.umich.edu/driverless-shuttle-service-coming-to-u-m-s-north-campus/

[12] Low Cost Shipping with AVs http://news.mit.edu/2016/driverless-truck-platoons-save-time-fuel-1221

# Chapter 5

# AV Plus Functional Component Design

*When technology challengers like Google and Tesla shake up a hundred-year-old auto industry that is very large and complex with a strong dose of computer hardware and software, it' is not easy to come up with an elegant architecture and component design. Yet it must be done right so that future add-ons do not interfere with the functional parts of the car. To automotive engineers using customized IC components, structured computer system principles and standards are quite foreign. How do we marry these disparate engineering disciplines in an AV Plus? It is not easy to implant an artificial brain into a physical body when it had a real human brain to drive it up until now. Nevertheless, it must be done in a clean and reliable fashion.*

In this chapter, I'll discuss some of the functional component design issues involved with autonomous vehicles. Let's start with the additional functionality we'll want in the AV compared to traditional non-AV vehicles.

## 5.1 Logical Functionality Required in a Fully Autonomous Vehicle

There are a number of cars available on the market today that are equipped with various ADAS features like adaptive cruise control, collision avoidance, lane-keeping assist, lane-change assist and parking assist. However, in order for a vehicle to be designated as fully autonomous (SAE level 5), it must meet the following minimum functional requirements:

a) Drive the car from point A to point B without the intervention of a human driver. It must perform this function observing all the applicable traffic rules and without getting into an accident. To achieve this, there are essentially three basic functions that an AV must provide—software-controlled steering, software-controlled throttling (up or down) to manage speed and software-controlled braking.
b) Respond to an owner's or passenger's call by bringing the parked car to a specified place, and park the car at a designated parking spot or any other vacant parking spot after the journey without the intervention of a human driver.
c) Connect to infotainment sources on the designated passenger's or AV owner's command.
d) Download maintenance/upgrade software over the air when required – a desirable feature (not listed in SAE specifications).
e) In designated emergency situations, hand over control to a human driver or bring the vehicle to safe stop at the nearest possible location. The industry is still debating who should be the ultimate boss: the AI software or the human driver.

f) In the case of a hardware, software or vehicle-component failure, move to a curb or safe position at the side of the road.
g) Perform all of these functions safely, observing all applicable traffic rules and without causing an accident.

## 5.2 Functional Components of an AV Plus

A quick note: In this book, I won't be discussing any of the traditional components of a vehicle (body, chassis, power source [internal combustion engine or battery], suspension, tires etc.) that don't directly contribute to making a car an AV Plus.

We can take two basic approaches to building an AV Plus:

1. Upgrade an ICE car to SAE levels 3–5 (in this book, I'm not interested in current implementations of SAE level-2 AVs) with smarter, connected, infotainment and autonomous hardware/software add-ons, including components like sensors and AI processors.
2. Upgrade an electric car to become SAE level 3–5 AV in a similar manner.

To upgrade either an ICE or an electric car, we need to add all or a subset of the following components (shown in Figure 5.1):

1. A number of sensors (simple object-detection sensors, more sophisticated LiDAR sensors, short-range/long-range radar sensors and cameras) and their connections to task-specific embedded controllers that feed the sensor information to a central computer
2. Sensor-fusion hardware/software that combines the input from multiple sensor sources into a holistic picture of the surroundings
3. The mechanical and electronic hardware and software features, including actuators, that control the three key driving components—steering, throttle control and braking
4. GPS and 3D maps
5. An artificial intelligence software engine—the brain of the autonomous car
6. A graphics processing unit or GPU (such as Nvidia's Drive Px platform) with high-speed processing capability and enough disk space to handle the huge amount of real-time data collected by the sensors and cameras
7. An interface to a communications network (DSRC [dedicated short-range communications] and/or 5G) for "connected-car" functionality, allowing it to connect with other cars in the vicinity and to the internet for mapping updates, traffic updates and infotainment
8. An interface to Apple CarPlay or Android Auto or similar future technologies
9. A user interface (UI) that allows the owner or authorized passenger to communicate with the AV

10. An operating system (OS) shell that manages all the hardware (computers and electronic interfaces to mechanical components), sensors and software components.

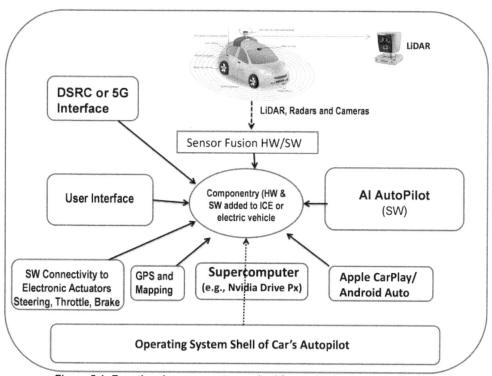

*Figure 5.1: Functional components required for an AV (logical and physical)*

Many of the components in Figure 5.1 are software-based; you won't physically see them in an AV. What you *will* see are the sensors and the supercomputer that runs the software. Figure 5.2 shows an AV with these components built in.

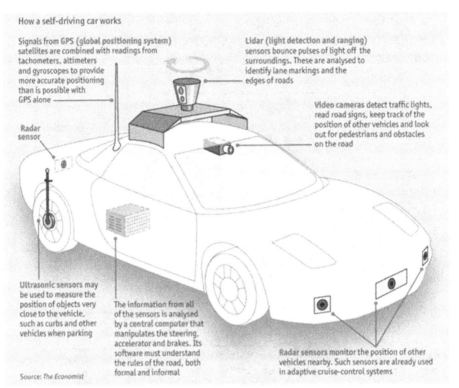

*Figure 5.2: Physical components of an AV (Source: The Economist)*

I'll first discuss the generic functions of the car's sensors and then look at how software interprets the information from these sensors to make meaningful decisions that drive the AV without a human.

## 5.3 How an AV "sees" its surroundings—Ultrasonic Sensors, LiDAR, Radar and Cameras

There are essentially three different types of AV sensors, as implemented in prototype AVs under trials in various countries: ultrasonic, LiDAR (Light Detection and Ranging) and radar. These sensors use different technologies, have different ranges of coverage and all provide a different set of information about the car's surroundings. AVs also use cameras, which I'll discuss later in this section.

**Ultrasonic sensors** use sound waves to detect objects in the vicinity of a car. A sensor sends out a sound wave at a certain frequency (as shown in Figure 5.3), and when the wave hits an object (a pedestrian, a car or a construction cone, for example), it bounces off the object and echoes back to the sensor. The nature of the echo indicates how far away the object is located. The most common application of this sensor today is in parking-assistance modules. Typically the range of these sensors is three to six meters, and their accuracy decreases with distance. To increase the range, engineers need to employ sophisticated signal processing chips. These sensors operate in the 40 kHz to 58 kHz bands.

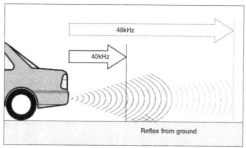

*Figure 5.3: Ultrasonic sensor*

**LiDAR** (shown in Figure 5.4) is the second type of sensor used in AVs; it uses light as a source of energy. It is the most visible sensor on an AV; you'll have noticed it on top of Google's and other self-driving cars (except for Teslas, which use cameras), with a rotating light beam going around in all directions. LiDAR is also the most important component that an AV uses to generate 3D images of its surroundings—determining where and how far away surrounding objects are located. The LiDAR system sends out a short light pulse that rebounds off an object such as the bumper or the hood of a car in front of it. The rebounded light pulse is detected by a receiver in the LiDAR unit, which then uses it to create a 3D image of an object.

*Figure 5.4: LiDAR sensor (Source: Velodyne)*

The typical range of a LiDAR sensor is 120–250 meters. The Velodyne LiDAR unit uses 64 laser beams and rotates 360 degrees, capturing information about objects around it in a 26-degree vertical field of view (FOV). The main issue with LiDAR is that it is quite expensive to make because some of the elements are made from rather expensive metals. In fact, the original Velodyne LiDAR used in Google's self-driving car was reported to cost in the $75,000 range. Newer models are much cheaper but are still too expensive for mass use. Several vendors are working hard to reduce this cost so that LiDAR can be used in millions of AVs in the future. Velodyne is an industry leader in this space; other players include Valeo and Leddartech. Velodyne itself has indicated that its R&D efforts into solid-state LiDAR could reduce the price to the $50–100 range per sensor.

LiDAR in automotive systems typically uses a 905-nanometer wavelength that can provide a range of up to 200 meters within a restricted FOV. Some companies are now marketing 1,550-nm LiDAR with longer range and higher accuracy, and are moving from mechanically rotating LiDARs to solid-state versions that are more reliable. LiDAR also requires optical filters to reduce the effects of ambient light. Velodyne claims that all of their LiDAR sensors are IEC class 1 laser products and as such are considered safe for human eye.

**Radar** (**Ra**dio Detection **a**nd **R**anging) has been around for ages and is used in a number of transportation applications including shipping, air travel and the on-the-ground public safety (police) sector. Radar can determine the velocity of a moving object (by using the "Doppler" effect), its distance away and the angle of the object. In AVs, radar sensors are mounted in the front and back to help detect surrounding objects.

Two key advantages of radar as a sensor are that it takes much less computer resources than other sensors and that it can work in almost all environmental conditions, including low light and bad weather. However, radar is unable to provide detailed qualitative information about an object it detects in any acceptable resolution—it can't tell whether an object is a pole or a person. Also, there's a trade-off between the radar's range and its FOV—the longer the range, the smaller the field of view. Long-range radar with a 15–30 degree FOV may see objects directly in front of it but not see objects in adjacent lanes of the road, especially in close proximity. Multi-beam radar solutions increase the FOV angle but are more expensive and increase the computing requirements.

Radar sensors can be classified by their operating ranges:

- Short-range radar (SRR): up to 30 meters
- Medium-range radar (MRR): 30–80 meters
- Long-range radar (LRR): 80–250 meters

LRR is the most popular sensor used in adaptive cruise control (ACC) and automatic emergency braking systems (AEBS) applications in the current generation of semi-autonomous vehicles. The current radar sensors for ACC and AEBS are unable to detect cars cutting across a vehicle from a lane on one side to a lane on the other. Future AV radar sensors will have to deal with this problem, as well as the issue of detecting thin-body vehicles such as motorcycles.

Currently, most radar sensors are two-dimensional, i.e., they send out waves on a horizontal plane, which means they can't detect the height of an object for which they detect the width. New 3D radar is under development to solve this problem.

**Cameras** (also called vision or imaging sensors): We've discussed three different types of sensors that use various types of waves to detect objects around a vehicle. Some people call cameras sensors, too—I prefer to categorize them as distinct devices, which are used to capture a 3D view of the surroundings. In fact, cameras have some clear advantages in that they can capture the color and the exact shape and profile of the objects they see. So, cameras are very useful for detecting things like traffic signals and changing lights.

Unfortunately, cameras, like the human eye, suffer some disadvantages as well. Cameras are adversely affected by bad weather and glare. They're not very good at night or under poor lighting conditions e.g. in tunnels. Sony has announced a special HDR camera that can see better in the dark, and lots of R&D is underway to help cameras see under difficult lighting conditions, but it's still work in progress. Nevertheless, cameras are the only sensor-like devices that can capture a faithful rendering of the surroundings as a car moves. These features, combined with ever-increasing pixel resolution and low price points, make cameras important components of a blended, multi-sensor solution. Figure 5.5[13] shows a typical arrangement of cameras for capturing detailed images of an AV's surroundings.

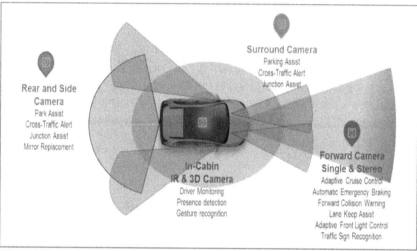

Figure 5.5: Typical camera configuration on an autonomous vehicle (Source: NXP)

The cameras of 2018 have a range of 120–150 meters. That range is expected to be extended to 250 meters in the next generation of cameras. Like radar sensors, cameras are already in use for various ADAS applications, but they still need continuous development to meet the needs of future AVs. Here are some examples:

- Adaptive Cruise Control (ACC): While today's cameras can detect full-width vehicles like cars and trucks, they eventually need to be able to detect thin-width vehicles such as motorcycles.
- Automatic High Beam Control (AHBC): Cameras currently enable high/low-beam switching and need to evolve to be able to detect oncoming vehicles and contour the headlight beams accordingly.
- Traffic Sign Recognition (TSR): Current systems recognize speed limits and a limited subset of signs. Future systems need to understand supplemental signs

and their context (such as "speed limit in effect 10am to 8pm"), and detect traffic signals and stop, slow down, etc. accordingly.
- Lane Keep Systems (LKS): These can currently detect lane markings, while future systems need to detect drivable surfaces, adapt to construction signs and account for multiple lane markings.
- Inside the vehicle, cameras will need to monitor the driver and passengers to enable gesture recognition and touchless controls. In SAE level-4 AVs, driver monitoring must take on the additional responsibility of determining whether the driver is sufficiently alert and aware to take control in the case of an emergency.

**Using the Communications Cloud to Enhance Information from Sensors and Cameras**

As far as I know, no standards have yet been defined for AVs' communication with the cloud. Nonetheless, while it's still very much in the conceptual stage, it's likely that we will be able to enhance the sensor information with real-time information from the cellular cloud. This would include information from multiple sources—other vehicles in the vicinity, emergency-response vehicles, transportation agencies (DOT), other drivers (like users of Google's WAZE), electronic traffic lights and yet-to-be-designed digital road signs, even if they are not directly within the car's line of sight. Eventually, when two or more vehicles come into range, they'll form an ad-hoc wireless network that allows them to transmit real-time safety-critical messages to each other.

## 5.4 Comparing Different Sensors

A great engineering debate is taking place in the industry as to the superiority of one sensor over another. Table 5.1 helps explain the key parameters and relative strengths and weaknesses of various sensors.

|  | Ultrasonic sensors | Radar | LiDAR | Cameras |
|---|---|---|---|---|
| **Basic technology** | Sound waves | Radio waves | Light pulses | Vision photography thr' illumination in the visible light spectrum |
| **Band** | 48 kHz | Approx. 3–30 MHz | Ultraviolet, visible, or near-infrared light; mostly uses 532 nm | Visible light spectrum |
| **Range** | 0–6 meters | Short: 30 m<br>Med: 30–80 m<br>Long: 250 m | 0–250 meters | 0–125 meters |
| **Key info element** | Presence of an object in the range—shape not identified | Presence of an object in the range—limited shape information | Pretty good but hazy picture of objects—not as clear as camera | Excellent image of objects and terrain—interpreted by computer vision |
| **Data intensity** | Very small amount of data | Small amount of data compared to LiDAR or camera; easy for a computer to process | Extremely large amount of data; computer-heavy | Huge amount of data; extremely high computer power required |
| **Strengths** | Low cost<br>Easy to implement | Master of motion detection<br>Proven technology<br>Works in all light and weather conditions | Adept at 3D mapping<br>Can get good profile of objects and surroundings | Master of capturing images as humans see<br>Most complete information of surroundings<br>Can be used for most AV applications |
| **Weaknesses** | Very short range<br>Slow speed | 2-dimensional, difficult to get a full 3D picture<br>Not enough detail | Very expensive | Works only in good light conditions |
| **Major applications** | Parking assist applications | Front and rear-view applications (collision assist, adaptive cruise, adaptive high/low beam) | 3D mapping applications (not cost-efficient for short-range apps) | Most applications including color-sensitive apps (street lights, road signs, speed signs) |
| **Cost range** | $15–20 | $50–100 short range; $125–150 long range | $100–8,000 — Costs will come down significantly | $125 for mono<br>$200 for stereo/3D |

*Table 5.1: Comparison of Sensors*

## 5.5 Current Cost Structure of Sensors and Cameras

Figure 5.6 shows the positioning and relative costs of the various AV sensors in 2017.

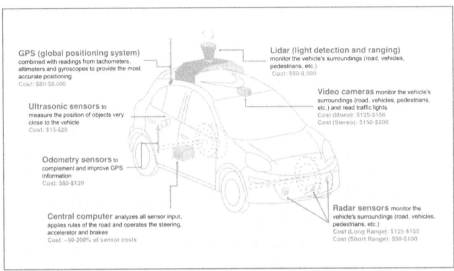

*Figure 5.6: Current cost structure of sensors (Source: Waymo)*

**Using All Four Technologies in an Optimal Fashion**

It is clear from the previous discussion that each of the three sensor technologies and the camera as an imaging device has distinct advantages, disadvantages and cost levels. Therefore, systems integrators use a combination of different sensors and cameras in various locations in the body of the car so that they can capture redundant information which is fused together to make a more composite picture of the surroundings. Cost optimization against a stated safety objective is another consideration in this engineering design and systems integration exercise – use the most cost effective and task-effective set of sensors for a given application.

## 5.6 The LiDAR versus Camera Debate (Waymo Versus Tesla)

Tesla and Waymo represent two sides of the LiDAR vs. camera debate in the AV industry. Tesla does not use LiDAR and relies on cameras instead for capturing 3D views of its cars' surroundings. On the other hand, Waymo uses all four technologies (ultrasonic sensors, LiDAR, radar and cameras) in their trial self-driving cars. Elon Musk of Tesla has come out publicly in favor of the "radar and camera only" approach versus the LiDAR, Radar and Camera" approach of Waymo. He appears to have based his decision primarily on the current cost structure of LiDAR sensors; he's motivated by his need to outfit his current set of cars with appropriate hardware at a reasonable cost so that he can upgrade them in the future (using future versions of his future version of "auto-pilot" software to upgrade those cars to what he calls fully autonomous cars. He also believes that the current configuration of sensors and cameras in the Tesla Model S will provide enough information about the surroundings that his (future) Autopilot software will be able to operate the car in full autonomous mode. Although he hasn't said so, to outside observers it appears that he's targeting SAE level 4 and not SAE level 5, where Waymo is headed.

My review of the two technologies and the problem they are trying to solve suggests that Tesla's approach may only offer a short-term solution. It may not be as safe in adverse weather and lighting conditions, because it relies on a human driver to be in control. Also in the future the suite of sensors will be cheaper and more capable, and will be able to handle situations and moving objects we haven't yet encountered.

In a 2016 Forbes article, Dr. Nidhi Kalra, a senior information scientist at RAND Corp. and co-director of its Center for Decision Making Under Uncertainty, is quoted as saying, "To choose now you're saying that today's technology will be the solution five years from now. The package of sensors you might want five years from now is probably going to be very different…. It's unconventional to think you can do this with cameras or radar alone. Cameras just aren't that good at detecting under all kinds conditions, and if you haven't trained a camera on a particular obstacle your perception software may not recognize that obstacle and know whether it's there or not, benign or not. LiDAR will tell you."

Another major consideration is that Waymo isn't concerned about the cost of LiDAR because the company is not yet building commercially available autonomous cars. Waymo believes that LiDAR costs will come down drastically and has patented LiDAR designs that it thinks can be produced for as little as $50–100, provided the production quantities are in the tens of millions of units—which is a reasonable assumption in my view. I don't see that a solid-state LiDAR should have a higher "bill of material" cost than $100. The scientists will figure out substitute materials for the rare metals that are currently required.

**Which Sensors Major AV Players Are Using in Their AV Prototypes and Trials**
Tesla's Model S is equipped with 12 long-range ultrasonic sensors that provide a 360-degree view of the objects around the car. Tesla also uses one forward-facing radar sensor and a suite of eight cameras, which are currently supplied by Mobileye. These cameras capture the location of the lane markers and the curvature of the road. Three front-facing cameras have the following range and FOV:

- Main forward camera: Max range 150 meters with a 50° FOV
- Narrow forward camera: Max range 250 meters with a 35° FOV
- Wide forward camera: Max range 60 meters with a 150° FOV

There are also cameras on the fenders and door panels, to capture side views, and a rear-view camera. The forward-facing radar plays an important role in getting information from further down the path the car is traveling because it has a greater range than the camera. Tesla uses a super computer from Nvidia (the Drive Px) to process the image data from all of the cameras, radar and ultrasonic sensors using end-to-end image-processing software called "Tesla Vision," which uses AI and neural networks. Drive Px is capable of performing 12 trillion operations per second.

**Waymo** uses all three types of sensors (ultrasonic short-range sensor, radar and its flagship LiDAR) as well as cameras for detecting objects and obtaining a picture of the car's surroundings. Figure 5.7 shows a Waymo self-driving configuration.

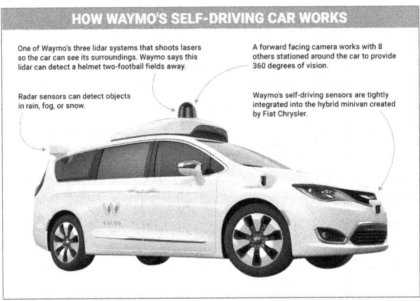

*Figure 5.7: How Waymo's self-driving car works (Source: Waymo)*

## 5.7 Sensor Fusion

Sensor fusion refers to the computer processing of information from different sensors and cameras to create a composite picture of the surroundings. This software is obviously quite sophisticated. In my view, the Waymo/Alphabet team has the best handle on this software as of 2018. Other companies are also trying to create versions for their own systems-integration efforts. This entails removing noise from the data that individual sensors provide and arbitrating between the information from multiple sensors (determining whether information from one sensor is more accurate than the same view from another sensor). Figure 5.8 from a Texas Instruments document written by Fernando Mujica[14], provides a good functional description of processes involved in the autonomous vehicle platform.

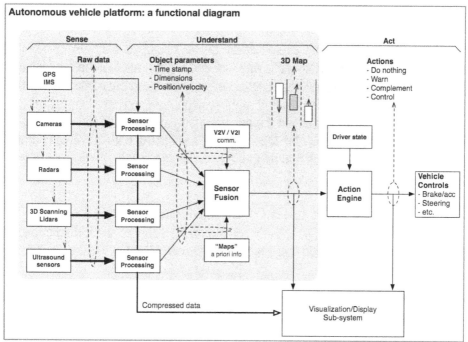

*Figure 5.8: A functional diagram of the autonomous vehicle (Source: Texas Instruments)*

## 5.8 Route Planning and Actual Driving[15]

*(Thanks to Shima Rayej of robohub.org for the information in this section)*

Each self-driving vehicle is equipped with not only the sensors and cameras described earlier but also a GPS unit and an inertial navigation unit (INS). An INS is a computer system based on three pieces of hardware – motion sensing hardware (accelerometer), rotation sensing hardware (gyroscope) and optionally magnetic sensor (magnetometer). It continuously calculates the position, orientation, and velocity (direction and speed of movement) of a moving object without the need for external references. The AV software builds a virtual 3D map of the road infrastructure and surroundings from the sensor and camera information and uses positional information from the GPS to locate itself on this virtual map.

The key management and execution for retrieving data from sensors, fusing that information into a meaningful and holistic picture and making decisions regarding the course of action to be taken is all done by supervisory software—what can be called the brain of AV. It is a very sophisticated piece of software and is still under continuous enhancement, refinement and development. I'll comment on the maturity of this software and the entire development process in later chapters but suffice it to say that, as of today, this software performs following tasks:

1. **Perception**: Maintain an internal conceptual map of its world (i.e., surroundings) and localize the AV on this map
2. **Decision and Control**: Develop a path to the AV's destination using the map but avoiding all the obstacles in the way and obeying the rules of the road including speed signs. The algorithms that this software uses are based on many simulations and real-life driving through which the AI software learns how to behave. This is what is called the deep-learning discipline of AI.
3. **Vehicle Manipulation** - Once the software selects the path, it converts that decision into commands that go to the actuators for steering, throttle and braking components.

These three steps are repeated thousands of times per second on the AV's supercomputer as fresh data is constantly coming off the sensors and terrain changes from origin to the final destination. Let's look at the process involved in each of these three steps in greater detail.

**Mapping and Localization**

As I've said, the AV's software must first develop a composite 3D map of its surroundings from the data it receives from sensors and cameras. The map may also use previously collected data about the surrounding terrain—Waymo uses this technique, by sending its human drivers on a route before the AV goes on its own. The previous 3D map data can take care of stationary objects and dynamic route sensor data can insert information about pedestrians, any new obstructions or construction and of course the actual traffic. Then it inserts its own location within that map—where it is at any given moment. It uses GPS, the inertial navigation unit and sensor data to determine its location on the dynamic map. This process is called localization. As Figure 5.9 shows, the picture of the surroundings assembled by the sensors and used by the Waymo car is pretty good but not close to the one seen by the human eye.

*Figure 5.9: The Google car's internal map at an intersection (Source: Waymo)*

## Obstacle Avoidance

The internal map used by an AV shows lane markings, other cars, pedestrians, buildings, traffic lights and stop signs. Obstacle information is compared with a library of obstacles, and the AV software makes a probabilistic determination of the direction and velocity of other vehicles and objects in the vicinity. While this is quite accurate on highways, it does present a challenge at intersections. The map shows the location of obstacles in three time periods: previous, present, and future, based on the speed at which they are moving. This information is subsequently used for path planning.

## Path Planning

The main objective of path planning is to use the information contained in the AV's map to direct it to its destination by observing the rules of the road and avoiding obstacles. While the specific algorithms used by different AV vendors are proprietary, they all use a general process that involves developing a long-range route plan and then a short-range path plan. The short-range plan may say something like, "drive forward in the same lane for 100 meters, change lanes because there's a slower vehicle up ahead and then turn right." This short-range path plan is re-calculated several times in a second. Once the path is determined, commands are sent to the actuators for the steering, throttle and braking subsystems.

## 5.9 Reducing Hardware Complexity Through Domain Controllers

Over the past few years, electronic engineers have been adding task-specific computers to the AV, resulting in 60 plus computers in a modern car. The new trend is to use domain and area controllers as depicted in Figure 5.11. This will reduce both the complexity and cost of AVs.

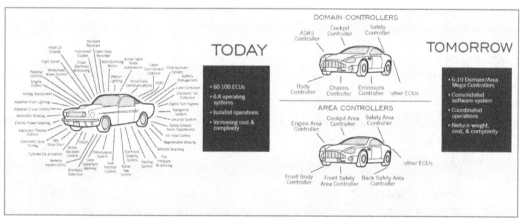

*Figure 5.10: Using domain controllers to reduce hardware complexity (Source: QNXauto.blogspot.com)*

## 5.10 The Shell for the Silicon Components of an AV—Operating System

The AV operating system (OS) software is the shell that ties together the distributed functionality of various sensors, handles data communications between electronic devices/embedded controllers, manages communication with vehicles and other nearby devices, assigns tasks to various application processes and then interfaces with electronic actuators for the three key tasks in driving the car (steering, throttling and braking).

The AV OS is an important piece of the AV. Should the PC desktop and smartphone history be repeated in the AV industry, decisions about the OS will influence the future direction and key innovations in this industry. I'll discuss this in detail in the next chapter.

## 5.11 The Supercomputer for AI processing

AVs process a huge amount of data in real-time. Processing of this data requires a very fast computer – no, a supercomputer. Companies like Nvidia and Intel/Mobileye are involved in developing these supercomputers. I shall talk about that in chapters 6 and 7.

### Summary

*In this chapter I have described various hardware components that are required to upgrade a non-AV to an AV. Most important of these components are the sensors of various types. The industry is still trying to arrive at the optimal configuration of sensors to do an effective job of "seeing" its surroundings. I have also explained how different sensors provide different type of information of the terrain and obstructions on the way. This disparate information from different is sensors is fused together to provide a holistic picture. The autopilot software then decides the path of the route and immediate movement of the vehicle to reach its destination. I shall discuss the role of software in the next chapter.*

---

### Citations for External References

[13] Cameras as Sensors – https://blog.nxp.com/wp-content/uploads/2017/05/Ali-Rear-and-Side-Camera.png

[14] Scalable Electronics for Autonomous Vehicles – A TI paper
http://www.ti.com/lit/wp/sszy010a/sszy010a.pdf

[15] How AVs work – Paper by Shima Rayej of robohub.org - https://robohub.org/how-do-self-driving-cars-work/

# Chapter 6

# The AV Plus OS, the Human-Machine Interface and AI

*A lot of attention is being paid to AV sensors, AI, TaaS, regulatory and infrastructure issues. These are important and fundamental to introducing this disruptive technology. However, there are three more pieces of the puzzle that are integral part of the picture I'm trying to paint: the human-machine interface (HMI), the operating software environment and AI. Let's now turn our attention to this trio, without which this picture is but a hazy cloud.*

## 6.1 Background

AV Plus cars will no longer be just 3,000-pound rolling machines made of steel, plastic and rubber with mechanical and electronic components. What will distinguish them from the current generation of vehicles is the extraordinary amount of computerization of all major components, the cars' use of battery power instead of internal combustion engines, their integration with smartphones, their rich infotainment applications and their network connectivity to other cars, to the internet cloud and to the city infrastructure. In today's semi-autonomous vehicles, there are upwards of 60 task-specific computers, as well as many advanced sensors (radar, LiDAR and software-controlled cameras) that provide information about the terrain and surroundings that is processed in real time by a supercomputer. The requirements of an advanced AV Plus are complex, demanding and mission-critical—literally protecting the lives of the vehicles' occupants. To ensure that autonomous cars are safer than human-driven cars, the future AV Plus must have a sophisticated and well-integrated operating system. The AV Plus OS will not only control, coordinate and manage the AI-driven self-driving function but also provide infotainment for enjoyment and monitor the health of the AV's organs in a cohesive and organized fashion. This will be a complex OS—far more complex than the OS controlling a smartphone or a desktop/laptop computer. Some experts have estimated that there could be 100 million lines of code in an AV—that's more than in a Boeing 787.

Whether the operating system manages the single cohesive brain of the AV or is simply a "hypervisor" loosely overseeing the distributed real and virtual machine environments of an AV is a moot point. I'll leave that question for the architects, designers and software engineers of Silicon Valley (or their proxies elsewhere). However, I do know that the future AV OS will be crucial to the success of AVs and that it will take a decade or longer to evolve and become stable. Until then, we may have to live with a variety of hastily put-together AV OSs as various manufacturers work to lead the way into the future.

## 6.2 Why the AV OS Is Extremely Important?

If we think of an AV as a human being, it must have a functioning physical body, a heart (an ignition engine or a battery) that gives it mobility and power, and a brain that gives it the intelligence to do what it's supposed to do, controlling its daily existence. Take out any of the three parts, and the AV is dead. The body and the heart of a vehicle have been around for decades and are now becoming safer, more functional and less polluting. But now vehicles are acquiring an AI-based brain that we hope will mimic a human one.

**I am suggesting that any vendor that supplies the OS, HMI and AI autopilot will control the AV development process to a large extent.** Whether there will be a small number of vendors or many, they will begin specifying the interface standards and providing reference design for the devices. OEMs will find that systems integration is easier if they follow the vendors' interface specifications. These vendors may even provide application development tools to help third parties develop AV-specific apps. Many of our current smartphone apps may be candidates for porting into the AV Plus platform; the first group being music, video and movie-streaming apps.

Many OEMs are trying to get in on the action, developing proprietary operating systems for their own AVs, but they'll soon find that the continuous development, maintenance and upgrade process for the entire life cycle of an AV requires a lot of financial and human resources. Gradually, only a small subset of vendors will survive. Look at other major industries; that is what happens. I expect the same phenomenon will happen in the AV industry.

In this fight for the technology leadership of the AV OS crown, I expect Waymo/Alphabet to survive because it has the technology expertise, financial resources and determination to pursue it for the long run. Also, the company doesn't have any vehicle-manufacturing ambitions that would make it a competitor to OEMs. Waymo can customize the OS for OEMs or the OEMs can customize it with Waymo's help. This is not unlike Google's modus operandi with smartphones. While Apple has been non-committal in this area, I expect that Apple could easily make its way into the AV industry and provide a very attractive OS with an intuitive user interface and strong AI that users would fall in love with. But Apple doesn't like to give its OS to others as a matter of control. QNX, a Blackberry subsidiary, has received a solid backing from many OEMs, including Ford. I'll talk about QNX a little later.

## 6.2 The AV Plus OS is Far More Complex than Most Computing Environments

In my 50 years' experience in the information technology industry across diverse situations, I have never come across an environment that is more complex than controlling the advanced AV Plus through an OS. Here's why it's so difficult and will require very sophisticated design, functionality and interfaces:

1. The AV Plus OS will have to manage a heterogeneous mix of standalone computers (which control low-level vehicle components), distributed computers and more powerful real-time computers all wrapped up together.
2. The amount of data that the AI supercomputer OS has to process is enormous, coming from sensors of different types. Wild guesses (and I think that these really are wild) have been made that AVs may generate 4,000 GB per day. I do not know whether this data has to be stored, where it has to be stored and for how long. If it's on the edge computer, it's manageable. If even a fraction of it has to be sent across the cloud via the 5G wireless network, that's an entirely different matter. Charges from the 5G service provider could add up to a huge recurring expense, which would threaten the entire AV business model. Who's going to pay for so much 5G wireless data?
3. Many of these distributed computer environments have nothing in common in terms of platform support because they're built by specialty vendors who are experts in a particular component, such as the vehicle transmission or the direct motor that supplies power to the wheels. The custom with these embedded computers is to use low-level machine-level code. These computers do not follow a common standard for external interfaces. Until standardization happens over the next few years, each component vendor will continue using their own custom interfaces.
4. Security is a key requirement in AV software architecture and has to be built into the fundamental design of the AV from the ground up—not afterward as a patch or module on top of the existing software.
5. The integration of different vendors' hardware and software is quite complex. The key reason that Apple is able to do it so elegantly for its iPhone and Mac platforms is that it's a vertically integrated company. The AV industry is vertically and horizontally distributed. Tesla is vertically integrated but Tesla doesn't have the bandwidth or the market acceptance to be the AV OS leader.
6. It's an AV OS jungle out there. Every OEM wants to cobble together its own unique OS. But in my view, the industry cannot sustain 15 different operating systems for very long. Consider how there are only two in the mature smartphone industry (iOS and Android), where there were six or seven in the late '90s. Similarly, there are only two operating systems in the laptop/desktop industry—Microsoft windows and Mac OS.
7. We need access to all of these computers through a unified and intuitive user interface—the medium for which has not been decided on or designed completely as yet. Whether we'll use the human voice, touch screens, a thumb wheel, human gestures or a combination of all four is still not clear. We'll need to interface at multiple levels: as technical service personnel, as private owners and as occupants of TaaS vehicles.
8. App development platforms have not become standardized in the AV industry. Everybody's using their own favorite platform. This does happen in the early days of a new technology. Apple's iOS app development platform is Swift. Android app developers use Java and AI app developers prefer Python. These platforms must co-exist. The chosen OS will have to deal with multiple platforms and insulate the apps

that use the AV's interface to interact with the human driver or the robo-driver in a clever way. It is apparent to me that display screen of choice for viewing is AV's display screen

9. Apart from the core driving task, the AV Plus OS must manage infotainment functions from disparate devices including smartphones using interfaces like Apple CarPlay and Android Auto. This integration is not standardized, although Blackberry's QNX umbrella software is currently leading the market and has the potential to become a defacto standard.

## 6.3 What Functions Must AV Plus OS Provide?

The OS for AV is essentially a software framework or logical superstructure that may perform some or all of the following important functions:

1. Act as a logical unifier of multitude of standalone or interdependent computers or domain controllers that reside in the car. The application function may be delegated to an individual computer but minimal supervisory control will still reside in the AV OS.
2. Create hooks into various in-car computers and sensors to provide two-way communication—input signals to the supercomputer (the central nervous system of the AV) and output commands from the supercomputer.
3. Manage a huge amount of data for the sensor-fusion application running within the supercomputer.
4. Interface with the GPS, the inter-navigational unit and 3D maps prepared by third parties or prior trips and also updated by real-time changes provided by cameras and sensors.
5. Interact with a human driver or his/her proxy through a user interface—exchanging information about what to do and where to go, accepting commands and calling for emergency help from a control center.
6. Support multiple types of user interfaces—touch, push button, thumb wheel, speech and gestures.
7. Provide protection against cybersecurity risks from individual computers such that any virus or hacking doesn't affect other critical components. This will be accomplished through a reverse hypervisor concept (interestingly, reverse of traditional hypervisor virtual machine concept — one hardware server with multiple machine incarnations) that separates rather than unifies the disparate computers.
8. Provide an environment for using in-car apps such as streaming entertainment, games, etc..
9. Manage the key application function of the autopilot driving mode.
10. Provide the ability to deal with all SAE levels of AVs, from level 3 to 5 (assuming that vehicles conforming to SAE levels 0–2 do not really need the OS we are talking about).
11. Interface with the infotainment modules of the car, including smartphone support through either services like Apple CarPlay or Android Auto.
12. Interface with vehicle-to-vehicle (V2V), vehicle-to-everything (V2X) and vehicle-to-infrastructure (V2I) communication modules through multiple network standards.

13. Interface with third-party AV subsystems and applications e.g. HERE HD mapping database and location services.
14. Provide OTA capability for updating software components.
15. Provide a mechanism for user profile settings (e.g., be able to offer different driving modes such as safe, normal or aggressive depending on the user).
16. Manage the interface with the internet cloud for updating the 3D maps in the car's supercomputer, obtaining real-time traffic and weather information and performing other proposed communication functions.
17. Grow with more functions as AV features evolve.

## Structure of a Typical AV OS

Different OS providers have their own architecture and block diagrams for their OS platforms. Figures 6.1 and 6.2 show two conceptual schematics of AV OS (one is generic and the other is from QNX).

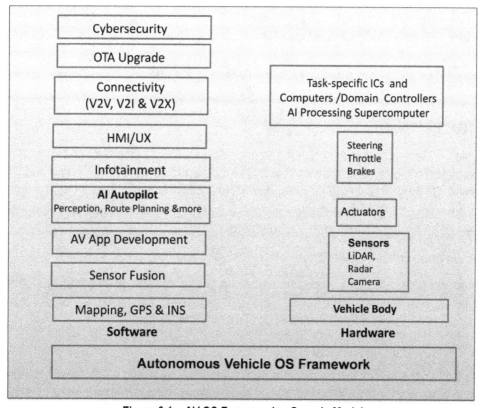

*Figure 6.1 – AV OS Framework – Generic Model*

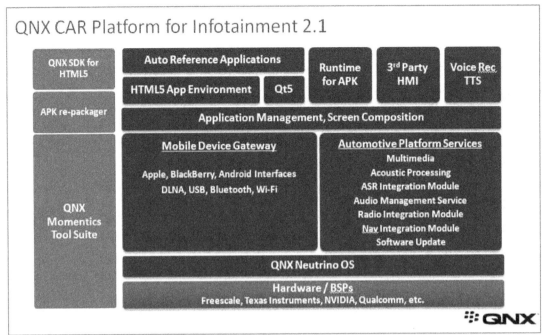

*Figure 6.2: QNX's platform diagram; TTS=Text To Speech, BSP=Board Support Package (Source: QNX)*

## 6.4 AV OS Players

The big OEMs realize that the OS will be critical to their AV development and monetization of apps, so they all want to build their own OS that they can brand and market it. Do they all have the expertise to do it well? Perhaps, not. Nonetheless, most of them have set up AV R&D centers in Silicon Valley and are putting together their unique notions of an AV OS for their own environments.

### 6.4.1 Waymo/Alphabet AV OS

Waymo has developed an AV operating environment for its self-driving car trials. I believe that the Waymo OS might be a strong candidate as a defacto standard, although Waymo has not talked publicly about wanting to license its OS to OEMs.

### 6.4.2 Tesla Autopilot as AV OS

Tesla is the only vendor, as of now, that has a vertically integrated AV OS that manages its suite of sensors, cameras, maps, AI decision-making criteria, route-planning and route-execution functions. The company calls its OS "Autopilot," and it consists of a hardware configuration and supporting software. Autopilot has gone through a number of revisions and is already on version 8 of its development cycle. Version 9 is expected in 2018.

**Will Tesla License Autopilot to Other OEMs?**

Industry observers thought that Tesla would keep its technology tightly under wraps, patented and out of reach from its competitors, but in 2017 Elon Musk proved them wrong. Tesla has now put all of its patents into the public domain and has invited anybody including its competitors to use its patents on electric cars—something that few technology visionaries have done. Musk has said that he wants the auto industry to switch to electric car technology and that if his patents help it happen faster, then he's very happy to share them in good faith. (See his blog post entitled "All Our Patent Are [sic] Belong To You" on the Tesla website.) [16] Whether he would do the same with his Autopilot software is a different matter. For security reasons, Musk won't open up Autopilot as yet. Besides, Tesla has a unique sensor configuration that differs (at this stage) with what most OEMs and Waymo are using—Tesla doesn't believe that AVs require LiDAR, and that camera-plus-radar-based sensor info is enough to capture the terrain of a route, while most other OEMs, as well as Waymo, think otherwise.

### 6.4.3 QNX

QNX, a Blackberry subsidiary based in Kanata, Canada, has become an important player in the AV OS arena. QNX claims that its Blackberry heritage gives it the required experience to build an OS for the real-time computing environment of an AV. QNX offers a broad portfolio of software solutions for the future of the automotive industry—cluster, infotainment, connectivity, in-car network security, type-1 hypervisor supporting multiple-ECU integration into domain controllers, ADAS and acoustics. QNX's OTA platform also provides a secure software upgrade capability. Add Blackberry's unmatched encryption technology to the mix and you have the ingredients for a hacker-proof operating system.

QNX software is a clear leader in providing software for managing infotainment, and is already being used in 56 million cars including brands such as Acura, Audi, BMW, Chrysler, Ford, GM, Honda, Hyundai, Jaguar, Land Rover, Maserati, Mercedes-Benz, Porsche, Toyota, and Volkswagen. An IHS Automotive report names QNX as the premier software supplier in the infotainment market, with more than 50% market share.

QNX's software platform provides several essential components that an OS must have. The following graphic from Blackberry/QNX shows these functions.

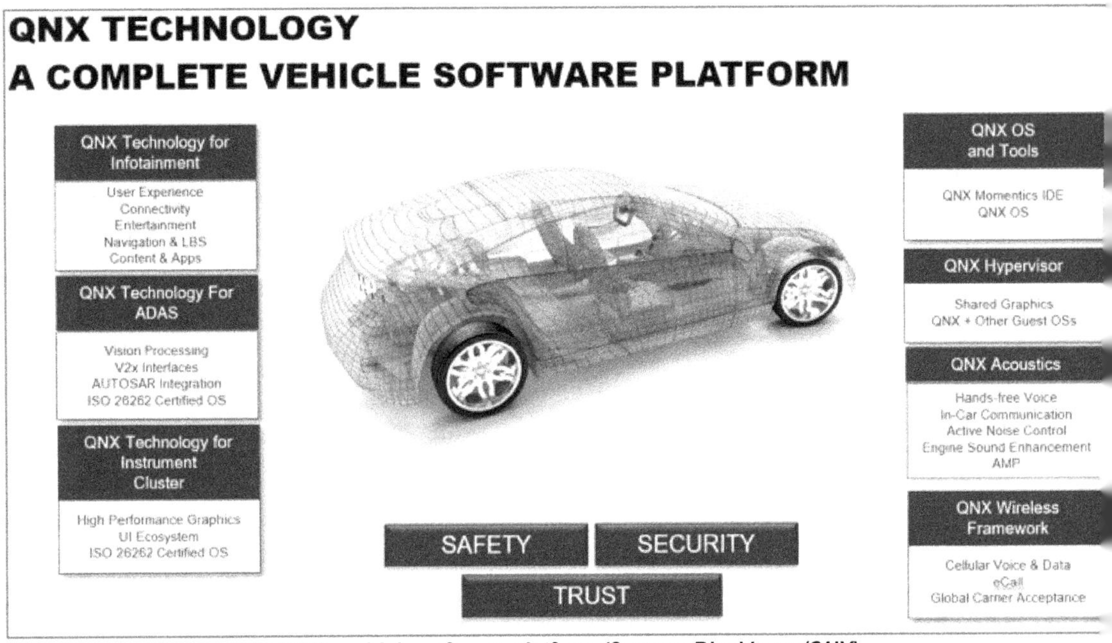

**QNX TECHNOLOGY**
**A COMPLETE VEHICLE SOFTWARE PLATFORM**

*QNX Technology for Infotainment*
- User Experience
- Connectivity
- Entertainment
- Navigation & LBS
- Content & Apps

*QNX Technology For ADAS*
- Vision Processing
- V2x Interfaces
- AUTOSAR Integration
- ISO 26262 Certified OS

*QNX Technology for Instrument Cluster*
- High Performance Graphics
- UI Ecosystem
- ISO 26262 Certified OS

SAFETY    SECURITY
TRUST

*QNX OS and Tools*
- QNX Momentics IDE
- QNX OS

*QNX Hypervisor*
- Shared Graphics
- QNX + Other Guest OSs

*QNX Acoustics*
- Hands-free Voice
- In-Car Communication
- Active Noise Control
- Engine Sound Enhancement
- AMP

*QNX Wireless Framework*
- Cellular Voice & Data
- eCall
- Global Carrier Acceptance

*Figure 6.3: QNX's vehicle software platform (Source: Blackberry/QNX)*

## Waymo Versus QNX

Waymo is deep into the self-driving-car technology business from a computing systems integration perspective and also as a TaaS provider. I consider Waymo a contender and leader for delivering an AV OS to the market since it has extensive experience in building operating systems for devices and it understands the AV Plus business very well—better than most OEMs, in fact. However, OEMs will want to make sure that they can use the Waymo OS, keep it under the hood and modify (add to/subtract from) it to give OEMs the brand differentiation that they may want. This might be tricky because Waymo cannot be an OS supplier and also be a competitor to OEMs. Waymo's current business focus appears to be on becoming a full-fledged TaaS player rather than on becoming an OS or autopilot supplier. QNX, on the other hand, is in different position. It is simply a provider of an OS framework that gives OEMs a lot of flexibility. OEMs can create their own version of the OS to provide the brand differentiation that they want. QNX is not a competitor to OEMs, and it may supply its OS to multiple OEMs.

### 6.4.4 Nvidia OS for AV

Nvidia has become an important hardware/software player in the AV landscape. The company has used its expertise in graphics-processing supercomputers, general-purpose OS development and AI to build software/hardware for autonomous cars. In fact, Nvidia claims that it has the fastest supercomputer for the AV marketplace. Nvidia hardware has already been used by several OEMs, giving the company an entry point into those OEMs' AV efforts. I consider Nvidia to be an important player in the AV OS landscape, including the most important part: the AI-based autopilot software.

Nvidia has approached the AV market by exploiting its strength in designing and building powerful computers and GPUs. It has developed a high-performance supercomputer to run the AI software, and it built the rest of the software (OS stacks and the autopilot AI) on top of its Drive PX Pegasus platform. On the other hand, the QNX OS approaches AVs from a software point of view and works with multiple hardware engines, including Nvidia's supercomputer running Nvidia's "autopilot" AI package.

### 6.4.5 Intel Mobileye OS for AV

In 2016, Intel purchased Israeli company Mobileye for over $15.3 billion for its expertise in building hardware sensors and "autopilot" AI. The company has every intention of offering a fully integrated hardware/software solution for the AV. Intel Mobileye has struck partnerships with several OEMs to supply parts of the solution as an appetizer now with a future objective of supplying the complete meal—including the dessert, as the company has its eye on sensors and LiDAR as well.

### 6.4.6 Tier-One Component Suppliers – Bosch, Delphi and Continental

Tier-one component suppliers Bosch, Delphi and Continental are also working on developing an OS framework that they can customize for an individual OEM.

### 6.4.7 Miscellaneous OS Suppliers ROS and Polysync[17] *(Credit for content - Polysync)*

**ROS** (which stands for Robot Operating System) is another AV operating system that some AV start-ups are using to build their self-driving car environments. My limited understanding of ROS suggests that it doesn't have the power and flexibility of an AV OS that a mission-critical, cybersecurity-protected environment requires.

Another start-up, **Polysync**, has made available an open-source kit of the AV OS they call OSCC (Open Source Car Control) that provides the core function of driving automation. OSCC allows developers to connect the OSCC modules, along with their own hardware and software, to the vehicle's internal control systems. This enables communication to the steering and throttle controls, using either the PolySync Core platform or some other software. Braking is enabled with the addition of a commonly available, repurposed automotive brake-by-wire module.

*Figure 6.4: An AV using Polysync's OS solution (Source: Polysync, Kia and DigitalTrends.com)*

Polysync's OSCC kit takes the mystery out of the automation function and provides a platform that educational/R&D institutions and start-ups can port into other vehicles that meet the wire-interface standards. In my opinion, however, it's not a full-function, production-level AV OS that OEMs could use.

Table 6.1 provides a comparison of the AV OS players.

|   |   | Key Features and Strengths | Weaknesses |
|---|---|---|---|
| 1. | Waymo | Strong in autopilot software, experience in OS design<br><br>Optimized for Waymo HW configuration<br><br>AV is Google's design, with body built by OEMs. | Not flexible enough to be ported to other OEMs' AVs, without significant modularization so that OEMs can pick and choose modules and brand it as their own OS. |
| 2. | Tesla | Highly integrated hardware and software configuration<br><br>Well-knit OS<br><br>Proven OTA solution | Not interested in selling OS to others, at this stage |
| 3. | QNX | Strong third-party OS framework that can be customized and branded by OEMs<br><br>Strong cybersecurity and security | Does not provide autopilot AI<br><br>Systems integration is complex |
| 4. | Nvidia | Well-integrated full stack of OS running on powerful Drive PX Pegasus HW platform | Only computing HW supported |
| 5. | Intel Mobileye | Very strong computing hardware heritage from Intel<br><br>Good potential | Not enough public information available |
| 6. | Bosch, Delphi and Continental | Rich auto component expertise gives them an advantage in systems integration | OS development is not core strength, but gradually getting better |
| 7. | Polysync and ROS | Open software stack<br><br>Good for educational institutes | Not a production-level OS |

*Table 6.1: Comparison of AV OS players; HW = hardware, OTA = Over The Air*

## 6.5 HMI (Human-Machine Interface) and UX (User Experience)

There are subtle differences between the user *interface* (another term for HMI) and the user *experience* (UX). While UX refers to all of the architectural, design and implementation aspects of a product that focus on its effective and enjoyable use by the user, the UI (or HMI) design is its compliment, the look and feel, the presentation and the interactivity with the product.

As the age of autonomous vehicles dawns, the human-machine interface and the user experience become extremely important. In fact, HMI and UX will influence the speed of adoption of AVs by consumers at large. If the user interface is easy and intuitive, the occupants/passengers of TaaS AVs will have a good experience. This will, of course, also apply to private owners. If the user interface is not easy, consumers won't likely want to give up their current generations of automobiles.

In traditional vehicles, the user interface has developed over several decades from the early days of static information display through dials, to today's limited interaction for changing lighting and entertainment settings. Today's user interface in non-AVs is fairly standard across manufacturers. That's one reason that most human drivers can get into any car and drive away, without any instructions. The accelerator is always on the right at the foot level, the brakes a little to the left. A handbrake is in one of two different locations. The console is pretty standard, too. Over the last 10–15 years, touch or non-touch digital screens with mouse wheels started appearing in cars, mainly for navigation, trip information, climate control and Bluetooth telephone interfaces. More recently (2016–2018), a smartphone interface (iPhone CarPlay or Android Auto) has been added to some models, allowing users to access their iPhone and Android applications including text messages and emails. Implementation of these digital interfaces is all over the map across OEMs—something users do not like as they switch from one car to another. This should be avoided with AVs.

Right now, most of the work on the AV HMI and UX is primarily investigative both by OEMs and by research firms that specialize in technology user interfaces and the experience discipline. This is a new area as far as AVs are concerned, and it's very complex and challenging. Like designing the AV technology itself, designing an acceptable HMI essentially involves a paradigm shift. The process that HMI specialty firms use is dependent on a reasonably clear understanding of the technology operation. The future operation of AVs is so much in a fog—what level of autonomy will be available and when, how occupants will sit in the vehicle, whether there will be pedals or not, what the medium of conversation with the autopilot will be, what the responsibility of the human driver will be and so on. All of these questions make the HMI very difficult to design and architect.

The difficulties that HMI designers face are many; let's discuss just a few.

### 6.5.1 Primary Methods of Interaction with Machines

Experts aren't yet clear on the ideal hardware and software configuration for the HMI. Questions abound: Who's the target user? Should we assume they're a computer-smart professional or an ordinary user who prefers to speak to the device in a conversational language of their choice? Tesla has chosen to install a 17" touch screen in its cars, while many high-end car manufacturers (such as Lexus and Mercedes) have taken away the touch screen and replaced it with a non-touch screen with a track-ball interface—is either of those the ideal device for the future AV? I do not think so. The ideal configuration will come over time as we try different implementations and users vote on it.

Should the device support human voice as the primary interface along with a touch screen? If yes, one would hope that the HMI will have a better understanding of each user's nuances and accent than Alexa or Google's Personal Assistant can handle today! Figure 6.5 summarizes the responses that Open Research Institute received when it asked how users wanted their AV to alert them to a deviation in its planned path.. ORI asked several other questions in its survey and HMI designers will have to research many similar questions and test prototype HMI based on consumer preferences. I am simply highlighting the basic issue.

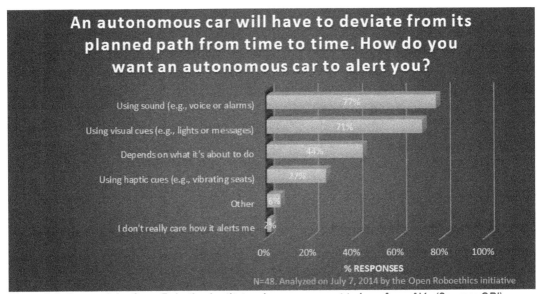

*Figure 6.5: Research by Open Robotics Institute – how users want to hear from AVs (Source: ORI)*

In addition to a conversational NLP interface, researchers in academia and advanced industrial labs are experimenting with hand gestures as additional methods of interfacing with AVs.

Another question for researchers and implementers to pursue: should there be two incarnations of the HMI – one for AV with pedals and one without?

How many permutations and combinations will the industry experiment with and force early users to struggle with before we settle on something that will work for most consumers across multiple generations? Could HMI personal-preference settings help solve this problem? I expect that it will inevitably take several iterations and a minimum of a decade before the dust settles and we have a de facto standard HMI. One thing's for sure: 15 plus OEMs and four or five technology vendors won't come to an agreement in such a competitive business environment; somebody (or multiple somebodies) will likely assume the HMI leadership, just as Apple and Google did in the smartphone industry.

### 6.5.2 Switching Between "Human" and "Autopilot" Mode

In AVs of SAE levels 2 to 4, we have either a human driver in control or an autopilot in control, and the ability to switch from one to the other. There's a lot of debate about when and how the "to and fro" conversation and switchover should take place. How should they alert each other? The protocol to be followed between a human and a machine is very important; a positive confirmation is required by each party to ensure a safe transfer. Should there just be a display message on the screen or an informational voice message as confirmation? What happens if the other party isn't ready to take over? What if the human driver is preoccupied with something? How much time should be allowed for the switchover? These are important considerations in HMI design.

### 6.5.3 HMI and UX Leadership

It's interesting to note that HMI and UX leadership doesn't necessarily reside with either the technology challengers or the OEMs. While technology challengers like Waymo/Alphabet may have superior software development capabilities, they may not fully understand drivers' expectations. OEMs have 50-plus years of experience in looking at the problem in the non-AV world. It's one thing to design an HMI for a smartphone, and quite another for an AV driving itself at 70 mph with four passengers reading the newspaper or surfing the web on those same phones! I believe that the only way to come up with a winning HMI is through a multi-disciplinary group of OEMs, technology challengers, HMI specialty firms and technology psychologists.

### 6.5.4 Learning from the Smartphone Industry

There is something that AV HMI architects can learn from the huge success of the smartphone industry. There's no doubt that the smartphone industry represents the largest population of technology users in the world. Also, the industry has achieved the fastest penetration of any consumer technology during the past century—from almost zero to 90% in the developed world and zero to 40%–50% across the world in under 15 years. I see four key reasons for this success:

I. The user interface is extremely easy and intuitive; even a two-year-old can learn it easily.
II. There are only two user interfaces and OS parents—Apple and Google—and even between the two, there's a lot of commonality.
III. 90% of apps are free (because developers benefit from ad clicks or views).
IV. App development interfaces are open and freely available.

### 6.5.5 Let HMI Designers Understand Testing Standards for Mission-Critical Technologies

Just as OEMs need to learn from technology challengers, computer industry enthusiasts must raise the bar they set for themselves in designing a user interface: with AVs, we can no longer send out beta versions of software that hasn't been tested thoroughly—the consequences of failure can be death. This is true especially for the case when a human driver has to take over from the autopilot in SAE level 2 to 4 AVs. Users can do weird things when under stress and asked to take over the driving task. Expectations should be unambiguous and choices few. The testing must be rigorous.

### 6.5.6 Customization, Preference Settings and Personalization

In a non-AV, very little customization and personalization is required. However, in an AV, designers will need to allow a user to customize the HMI extensively, allowing them to set preferences and/or train the system for things such as:

- Language
- Users' pronunciation and accent
- Types of driving—aggressive, regular or relatively slow and steady
- Default infotainment options;
- Navigational defaults;
- When to allow OTA upgrades (knowing that communication charges may be incurred)
- The types of data that can be collected (in view of privacy issues)
- Methods of alerting the human driver to take over

We can expect that AV HMI experts will conduct extensive user testing to learn consumer preferences. I cannot overemphasize the need for a simple HMI. An average AV consumer is not an airline pilot with years of flying experience. Nor are they sitting in the cockpit of an airplane with intense attention and focus.

Extensive customization and personalization options should be available so that users can set the preferences that they are comfortable with.

The HMI should recognize that some consumers may want to use a classic user interface—one that they're used to from their prior driving experiences. Although a voice interface may be available, some consumers may be more comfortable with the type of touch screen interface that's most prevalent in their smartphones. Once they're comfortable with the AV in terms of basic functionality, users may learn to trust the AV and graduate to a more natural voice interface (after training the NLP engine with their individual voice, accent and commands).

The industry will make huge progress once a conversational voice interface becomes available and autopilot can understand a variety of accents from regions of different countries. We're not there yet in current versions of voice commands in many luxury automobiles. I've had limited success with Mercedes's voice training program in my own car (Mercedes with intelligent drive). HMI testers will find that their voice engines do not understand the great variety of accents from different regions of the United States, English and French Canada and European countries. The ability to receive confirmation from the user that the machine has understood the human will be an important attribute of a good HMI.

### On-Demand User Interface Based on "Conceal and Anticipate" Principles[18]

In a 2015 article entitled "On Demand Dashboards," author Cobie Everdell suggests that the modern AV UI should present information only *as the driver needs it*, concealing information that is not required at a given moment and anticipating what the driver might need based on the situation. The user should still be able to retrieve other ongoing performance information it if he/she needs it, but overall the information presented should be exactly what the driver needs and anticipates in almost every situation. As an example, new navigation systems are being designed to advise the driver when to turn before they have to turn, with a command like, "please prepare to take the next exit on the right." This suggestion by Everdell seems to make sense and is being adopted by many implementers.

#### 6.5.7 Building Trust through HMI & UX

Industry experts have concluded, and I agree, that consumer trust will be one of the most important factors in determining the adoption rate of AVs in the initial two phases (as defined in Chapter 14). This trust will be highly dependent on the HMI and on the AV's ability to perform its driving task safely and accurately. The auto industry will have to use all of its marketing skills and persuasive expertise, offering free rides and giveaways and making extensive use of word-of-mouth reviews and other techniques, to win consumer trust. As of now, the majority of car-buying consumers (regardless of their generation) aren't ready to trust the current AV technology.

## 6.6 "Best of Breed" Autopilot for AV

If I had to assemble a "best of breed" autopilot, here are the ideal components and software stacks I would want to include, along with some additional requirements that may represent the wishful thinking of an AV OS purist.

- QNX's framework for infotainment, cluster control and cybersecurity based on Blackberry's well-known encryption technology
- Nvidia's supercomputer platform
- Middleware that insulates the higher layers of the OS as consolidation continues from 60-odd computers to smaller numbers of domain controllers
- Support for redundant and fail-safe architecture across critical components
- Waymo's AI software engine for sensor input, sensor fusion, route planning and algorithmic execution
- HERE's HD 3D-mapping and location services
- Apple's standard for an easy-to-use, intuitive human-machine interface or an interface using yet-to-be-defined human-like natural language processing (NLP)
- Tesla's user experience
- Android's "open" application development
- Apple CarPlay and Android Auto interface support
- NHTSA SAE-level support—ability to upgrade from level 2 to level 5 as the industry moves forward

- GM's OnStar platform for over-the-cloud telematics and user support
- Tesla or iPhone/Android OTA management subsystem for upgrading software over the cloud

## 6.7 AI in AVs

No Autonomous Vehicle book would be complete without at least a summary discussion of the role AI plays in eliminating the need for a human driver. Therefore we are providing a brief and simple explanation of AI in the AV context. Let's first look at what AI is in general terms and then at how it's applied in an AV.

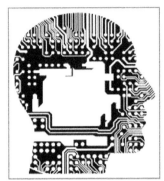

*Figure 6.6 – AI - Credit openclipart.org*

### 6.7.1 What is AI?

Artificial Intelligence refers to computer-based technology that gives human-like intelligence to computer applications, going above and beyond traditional computer applications that essentially calculate, search databases, retrieve information and perform rule-based tasks. When the answer to a specific question or a response to a specific request is deterministic and no intelligence is required (i.e., when there's only one answer), an AI application is not required. However, if the task is complicated, as in a medical diagnosis based on a certain combination of symptoms or a prediction that a machine will fail based on a certain pattern of results, where software is trained to learn from and interpret large amounts of data, it may require an AI application. Similarly, when autonomous-car software plans a route and drives the car after taking into account the information about the terrain and all the other cars and pedestrians around it, constructed from the digital input from sensors and cameras, then that advanced computer application is using the rapidly evolving discipline called AI. The key difference here is that computer software learns from the data it is fed—autonomous cars learn just as human drivers do when they're guided by a driving instructor or a more experienced human driver, and even when driving alone in a multitude of situations.

The relationship between the data that we feed about past experiences in the same context and the predictions (result of a requested investigation) the AI software makes is not linear, deterministic and well-defined using good old "if-then" computer logic rules. AI algorithms make decisions or predictions based on recognizing, as an example, a combination of pixels of an image as dog's face and then through another pass of the same image confirming that this is most likely a dog. I shall describe this methodology a little later as to how AI does this through training – looking at the data with some external guidance. When traditional voice-recognition software responds with a specific answer to a fairly structured query, that's not AI. But when the software can understand the intent of the question through a variable conversation that is not highly structured, it is intelligent interpretation on the part of the software and it would fall into the realm of AI. That is how modern customer service chatbots work.

### 6.7.2 Machine Learning and Deep Learning

Let's now turn to machine-learning and deep-learning concepts within the general discipline of artificial intelligence. I'd like to credit several sources for this information. In particular, Roger Parloff''s 2016 feature article published in Fortune magazine[19] (http://fortune.com/ai-artificial-intelligence-deep-machine-learning/), David Kelnar's 2016 *Medium* article[20] "The fourth industrial revolution: a primer on Artificial Intelligence (AI)" and Jay Jacobs' 2017 *Barron's* article[21], "Artificial Intelligence, Explained."

Machine learning is a subset of AI, and deep learning is a further subset of machine learning, as illustrated in Figure 6.6.

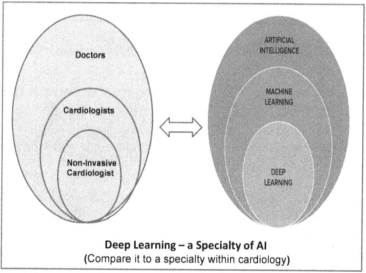

*Figure 6.7: Deep learning as a specialty within AI (Source: Barron's)*

Machine learning focuses on AI computer systems that can learn from data, rather than being explicitly programmed to perform tasks based on pre-defined rules. Machine learning analyzes data it is fed, discovers patterns in the data and uses those patterns to make predictions. The AI machine is essentially being trained to perform a task more accurately as it learns, building its own logic and solutions. This is similar to the way young children become wiser and wiser as they observe the actions of others.

Machine learning can be supervised or unsupervised. In supervised learning, an algorithm receives input and the likely outcome. Then the machine is allowed to figure out the relationship between the two. The algorithm notes any difference between the prediction and the correct output, then it tunes the weighting of the input variables to improve its future accuracy. In unsupervised learning, random data for a certain situation is provided and the machine is asked to find a relationship and a solution.

There are many approaches to machine learning. The **random forests approach** creates multitudes of decision trees to optimize a prediction. The **Bayesian networks approach** (named after probability mathematician Thomas Bayes) uses a probabilistic approach to variables and the relationship between them. **Support vector machines** are categorized examples and create models to assign new inputs to one of the categories. But none of these approaches solve the problem of autonomous vehicles. Cars come in too many shapes, sizes and colors. There are too many variations of cars and too many obstructions to deal with while driving.

**Deep learning** is a special approach within the machine-learning discipline of AI in which a computer makes decisions and essentially *learns on its own*, with no feature specification or optimization by the programmer. I'll explain this in more detail shortly. I should note that deep learning is not required for solving all problems. It's a highly complex technology, needs a lot of data, requires considerable analytical work by AI experts and consumes a huge amount of computing power. There is a lot of matrix manipulation that requires specialized graphics processors with thousands of cores, unlike a desktop computer that has only 4-8 cores

### 6.7.3 Choosing a Sweet Mango—Making AI and Machine-Learning Concepts Simpler to Understand

In searching for a simpler explanation of AI and machine-learning concepts, I came across a nice example by Pararth Shah of Stanford University on bigdata-madesimple.com website. In an article entitled "How Do You Explain Machine Learning and Data Mining to a Layman?" Shah helps the uninitiated to understand these wonderful concepts through the following simple example.

Let's say you're shopping for mangoes, and you want to buy mangoes that are sweet. However you do not know how to select sweet mangoes. Your mother once told you that bright yellow mangoes are sweet, so you buy those from your fruit vendor. Unfortunately, when you cut open the mangoes at home, some of them aren't sweet; in fact, only half the mangoes are sweet. So you conclude that your mother is not right all the time.

Your grandmother once told you that bright red and pulpy mangoes are juicy and sweet. Now you try that kind of mango, and again meet with mixed results. The next time you go to the local market, you find that your fruit vendor has gone out of business. You look for another vendor who sells mangoes, and find one who sells mangoes from Mexico. Mexican mangoes have entirely different color and taste than those of the previous vendor. As you gain experience with the taste of mangoes and the different variables involved—size, color, place of origin, softness of pulp etc.—you start wondering whether you can codify your research and develop a relatively reliable method of predicting the sweetness of any mango. But you come to the conclusion that there are too many variables; you cannot come up with a reliable mango-selection methodology.

Confused and perplexed, you meet a friend from Silicon Valley who's a traditional software programmer. He assures you that he can write an iPhone app that will figure out how to pick sweet mangoes. He asks you to write all the rules for selecting the right mango. You try as best you can with the knowledge you have. He starts writing "if-then" logic, but the logic becomes very complex, like the roots of a tree. Eventually he gives you the iPhone app. You try it but it does not work well because the conditions have changed in the market.

You go back to your programmer friend and complain. He concedes that the problem is more complex than he thought and suggests that he will refer you to a computer brain neurologist from Silicon Valley who is a brilliant AI scientist specializing in fruit technology. Mangoes are your eating passion and AI applied to fruits is his. You pose your dilemma to him and he suggests that he build a machine-learning algorithm for mango selection. Here is his solution:

Take a randomly selected group of mangoes from the market; call this training data. Make a table of all the characteristics of each mango like color, size, shape, ripeness, the country where it's grown, the vendor it's sold by, its pulpiness, etc. (these are your **input variables**) as well as their sweetness and juiciness (your **output variables**). Feed this data into a machine-learning algorithm. Out of various algorithm types, let us pick classification/regression. This algorithm learns the correlation between an average mango's physical characteristics and its quality.

The next time you go to the market, you measure the characteristics of the mangoes on sale (this is your **test data**) and feed it into your algorithm. (The mango vendor looks at you and wonders what this mango aficionado is doing with his iPhone and his mangoes with no intention of buying.) You are happy with the initial results and, more importantly, the results get better every time you use it. Your AI mango-selection algorithm uses rules similar to the rules you initially wrote out manually, but you don't need to worry about them anymore. The system is smart enough to create better rules as conditions change. You decide to become an AI fruit specialist and start developing apps for apples, oranges, bananas, grapes and more. You sell these apps to fruit buyers and, naturally, make millions.

### 6.7.4 Deep Learning and Neural Networks

Now you understand basic machine learning concepts a bit better. But we are not done with AI. You must also understand *deep learning* and *neural networks* if you're going to hold your own among AI and AV experts.

As I said, **deep learning** is a subset of machine learning in which a computer essentially *learns on its own*. How is this done? This breakthrough in deep learning came when computer scientists started modeling the brain rather than the world. Human brains learn to do complex things like understanding speech and recognizing objects (simple things that are done sub-consciously) through practice and feedback. Young children are taught this way. To do this, the human brain contains millions of neurons that are connected through neuron networks that grow and change as we learn. So computer scientists started mimicking human neuron networks and called them **neural networks**—now a common term in AI.

Deep learning is what gives autonomous vehicles the software brains that can drive without a human driver. AI calculators simulate the function of the human brain's neurons in a computer neural network. The neural network receives an input (e.g., an image of a car, a pedestrian or a dog), analyzes it and makes a prediction about what it sees. This output prediction is compared with the correct output and feedback is provided to the neural network. When the prediction is incorrect, the connection between neurons is adjusted by the network in order to avoid making the same mistake again. This process continues with more and more data till the system reaches a stable situation in which the result is correct most of the time. So practicing with more and more data makes our AI engine (a human-brain simulator, so to speak) almost perfect most of the time. **Note my use of the words** *almost* **and** *most of the time.* A lot of training is required before the system can accept data (for testing, validation and further learning).

According to Kelnar, "a neural network is created when AI neurons are connected to one another; the output of one neuron becomes an input for another layer. Neural networks are organized into multiple layers of neurons. The 'input layer' receives information the network will process—for example, a set of pictures. The 'output layer' provides the results. Between the input and output layers are 'hidden layers' where most activity occurs. Typically, the outputs of each neuron on one level of the neural network serve as the inputs for each of the neurons in the next layer".

Let's look at how an AV would recognize a dog as an obstacle, using an illustration from Parloff's *Fortune* magazine article. When data is fed into a neural network, the network looks first at low-level differences on the edges or outlines of an image and determines whether they fit its "learned" knowledge of dog images. As the image goes to the next layer of the neural network, the system compares higher-level features of the images, such as the feet and face of the dog. This carries on till the system has processed every layer and confirmed that the image fits its understanding of dog images at all layers.

Using this technology, AV systems can recognize pedestrians, dogs and other moving cars on the road. At its output layer, based on its training, the neural network will decide with a certain degree of probability that the picture is of the specified type (a dog with 95% probability).

### 6.7.5 Which Autonomous Driving Functions Require AI?

Driving is a complex human activity—as we drive, we are constantly involved in observing traffic rules, watching the traffic, looking out for pedestrians, recognizing the terrain, allowing for hazards and obstructions, knowing the map, figuring out the route and keeping ourselves fully alert. Robotic autopilot is, by far, the most complex AI application that scientists are trying to develop today. As several experts have concluded, it is equivalent in complexity to Kennedy's moon-shot project. In 2017, Apple CEO Tim Cook told Bloomberg, "We sort of see it as the mother all AI projects. It's probably one of the most difficult AI projects to work on."

## 6.7.6 What is Next in AI and Deep Learning for AV?

We think we still have a long way to go to make our AI autopilot perfect. More research and much more test data will be required. It's not enough to run various AI algorithms through only 10 million (2018) miles of real-life driving data. Recent accidents with Tesla and Uber AV trials have left the uncomfortable feeling that we don't fully appreciate the amount of testing required to confidently eliminate the human driver. We may need 10–20 times more test data before we will all be comfortable with a mission-critical autopilot.

As Professor Hinton and other leading AI researchers have indicated, we would need more advanced AI methodologies beyond deep learning and neural networks to truly (lets us say, in 99.99% situations) mimic human driver's brain while driving in terms of logically viewing surroundings, perceiving and making decisions.

## Summary
*In this chapter, I have emphasized the importance of the AV OS and HMI as key factors for the adoption of AVs. Equally important is the need for more research and development into AI algorithms for AVs. The industry also needs additional driving data to validate AI algorithms and the autopilot driving decisions based on them.*

---

### Citations for External References

[16] Tesla Blogpost Regarding its Patents - https://www.tesla.com/blog/all-our-patent-are-belong-you?redirect=no

[17] Polysync OS – https://www.polysync.com

[18] On Demand Dashboards - https://designmind.frogdesign.com/2015/10/on-demand-dashboards/

[19] Fortune (Sep 2016) Article by Roger Parloff - http://fortune.com/ai-artificial-intelligence-deep-machine-learning/

[20] David Kelnar's article on AI - The fourth industrial revolution: a primer on Artificial Intelligence - https://medium.com/mmc-writes/the-fourth-industrial-revolution-a-primer-on-artificial-intelligence-ai-ff5e7fffcae1

[21] Jay Jacobs article "AI Explained"- https://www.barrons.com/articles/sponsored/artificial-intelligence-explained-1508530169

# Chapter 7

# The AV Plus Ecosystem

> *The automobile industry is very old and very large, with an ecosystem that is quite diverse and extensive. It encompasses traditional OEMs, tier-1 component suppliers, the dealer service network and multiple other companies. With the emerging AV Plus industry disrupting the traditional auto industry, hundreds of new vendors are joining this ecosystem, and they all have their own strategies and plans. We need to understand this ecosystem so that we can analyze the impact of the changes that the AV Plus industry will bring about. Tomorrow's auto ecosystem will be different. Those willing to change and adapt will survive—and others will be wiped out.*

The automotive industry is, by far, the largest integrated manufacturing industry in North America and Europe. It has a very large, interconnected ecosystem with strong vertical integration and well-formed relationships between established OEMs and a network of tier-1 suppliers that provide various components to the OEMs. With the arrival of autonomous vehicles, the ecosystem is growing even larger as AVs exploit innovative forces of the traditional automotive centers (Detroit/Oshawa/Stuttgart/Munich/Mexico) and new entrants from Silicon Valley in Califonia, Ottawa in Canada and Munich in Germany.

We can categorize AV Plus ecosystem members into eight distinct categories:

1. Technology newcomers: Tesla, Waymo/Alphabet and Apple
2. Established OEMs: GM, Ford, Volvo, Daimler, BMW, Audi, Toyota, FCA, Hyundai, etc.
3. "Me too" AV players: Didi, NIO
4. Major tier-1 suppliers: Bosch, Continental, Delphi, Magna
5. Computer-centric AV component players: Intel/Mobileye, Texas Instruments, NXP, Nvidia
6. AV Plus OS and HMI players: QNX/Blackberry
7. AV specialty players: Drive.ai
8. TaaS players: Uber, Lyft, Baidu (China), Ola (India), and others
9. AV fleet vendors: Navya etc.

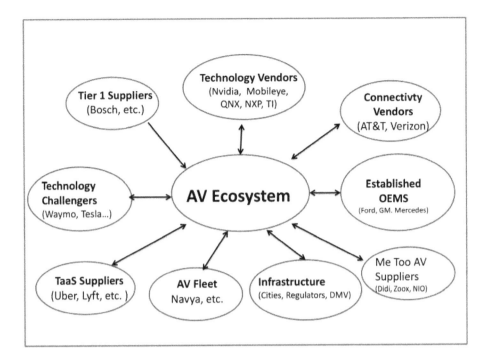

In this chapter, we'll look at the business strategy and current status of the first seven categories; we'll discuss the last two in Chapter 8.

## 7.1 Technology Newcomers

### 7.1.1 Waymo/Google

In the judgment of many analysts, including me, Waymo/Google can be credited with crystallizing the AV commercialization process in the recent past. In spite of a lot of R&D done in Europe and the United States over the past 80 years, it wasn't until Google started its self-driving-vehicle project that the idea captured the attention of automakers, innovators and the public alike. German automakers like Daimler had done excellent development work with academic partners in 1980–90, but they remained "R&D" partners with no concrete plans to commercialize their work. What Google did was to light a spark under self-driving car innovation. The company's efforts were certainly bolstered by its marketing finesse in allowing the public to see its self-driving car on the road, in Silicon Valley where innovations get noticed. And Google's credibility in the technology-focused world made people start believing that this was something big.

*Figure 7.1 – Google's self-driving Firefly (Source: Waymo)*

Google's self-driving project was initiated by Google's founder, Sergey Brin, and was led by Sebastian Thrun, who has been called the father of the AV commercialization process. However, I think the credit must also go to a number of researchers in Germany, at CMU, at Stanford and in Parma, Italy, whose efforts enabled Google's trials.

Some important points about Waymo's development:

- The self-driving-car project at Google (now Waymo, a subsidiary of Alphabet) formally started in 2009. Actual work had begun a bit earlier in stealth mode when Thrun, a professor from Stanford University who had participated in the DARPA Grand Challenge in 2005, became an advisor to Google.
- The project began with six Toyota Prius cars and one Audi TT model, and later switched to a Lexus RX 450h. The company also built a custom self-driving car and later built its well-known Firefly model, which has been in the media all over the world. Waymo also started "Drive Me" trials using Chrysler Pacifica minivans in Phoenix, Arizona, in 2017.
- By the end of February 2018, Waymo's self-driving prototype vehicles had already logged over five million actual miles on private and public roads and over a billion miles in simulation. Google's self-driving cars rack up three million simulated miles every day in its giant data centers. At any time, there are now 25,000 virtual self-driving cars making their (computer-simulated) way through fully modeled versions of Austin, Mountain View, and Phoenix, as well as test-track scenarios. You can learn more about this in Alexis C. Madrigal's 2017 *Atlantic* article "Inside Waymo's Secret World for Training Self-Driving Cars." [22]
- Waymo has patented and designed prototypes of inexpensive LiDAR, radar sensors and cameras.
- Waymo has more complete engineering experience in building an OS for AVs than any other company.

- Waymo has extensive experience in AI deep learning—the heart of the AV—perhaps more than any of its competitors.
- Waymo has built a prototype car without a steering wheel or accelerator—essentially a concept car for an SAE level 4 or 5 fully autonomous vehicle.
- The key difference between Tesla's and Waymo's experiences is that Tesla has more cars and more real car miles on the road, but most of those miles are being driven by human drivers operating in SAE level-2 or level-3 AV mode. Waymo's experience base is much closer to level-4 or level-5 AVs.

**Waymo's AV Strategy**

To outside analysts like me, it appears that Waymo is keeping its marketing strategy quite open right now. While it is true that it hasn't yet converted any of its R&D into a revenue-generating product or service, it does have a lot of options for entering the market as it matures. Nobody's making any revenue from AVs so far anyway. Waymo's goals are like a moon landing—build products for a fully autonomous SAE level-5 car, not a level 3. Because of its strong market position and huge bankrolling ability to acquire exceptional talent or credible start-ups, Waymo could pursue any one or a multiple of the following options:

- License its software (OS and other key components) to OEMs: strong probability
- License its sensor designs—solid-state LiDAR, radar and cameras—to tier-1 AV component suppliers: strong probability
- Increase its stake in ride-hailing vendors like Lyft—the company led a $1.5-billion funding commitment to Lyft in late 2017
- Invest in or acquire or form a strategic partnership with an established OEM: medium probability
- Convert Waymo into a full-fledged vertically integrated car company like Tesla: unlikely

Irrespective of what path it follows, Waymo will be a formidable player in the AV ecosystem.

### 7.1.2 Tesla

Tesla, of course, is another technology challenger in the AV commercialization race. I may give Waymo slightly more credit in the AV commercialization effort, but Tesla had earned an admiration of the technology professionals for producing an all-electric luxury car from scratch using mostly Tesla-designed components and Autopilot software coded by Tesla engineers. These consumers love what Tesla has so far produced: a luxury Model S sedan, an SUV, a mass consumer Model 3 and soon a truck—all electric and all with highest mileage range in the industry. Investors have voted with confidence—making Tesla's stock valuation higher than that of GM and Ford on certain days in 2017. Whether it's irrational exuberance or not and whether it can be sustained, Tesla has won the confidence of the upper echelons of the American professional class.

Tesla was founded in 2003 by a group of engineers who wanted to prove that people didn't need to compromise to drive electric cars—that electric vehicles can be better, quicker and more fun to drive than gasoline-powered cars. Tesla designed the world's first premium all-electric sedan—the Model S— from the ground up, and it has become a much-talked-about car among technology-aware professionals. The Model S has the longest range of any electric vehicle, over-the-air (OTA) software updates that make it better over time, and a record 0–60 mph acceleration time of 2.28 seconds (as measured by Motor Trend). In 2015, Tesla expanded its product line with the Model X, a sport utility vehicle that holds 5-star safety ratings across every category from the NHTSA. Completing CEO Elon Musk's "Secret Master Plan," in 2016 Tesla announced the Model 3, a low-priced ($35,000), high-volume electric vehicle that began modest production in 2017. Model 3 has run into production problems that had not been fully resolved at the time of printing.

Musk expects fully autonomous Teslas to be ready by 2019, but he notes that regulatory approval may take one to three more years to secure. He's still talking about running an autonomous trial run from his Los Angeles parking lot to a New York parking lot by mid-2018.

## Comparing Google/Waymo's and Tesla's Strategies

The following table provides a quick comparison of the strengths and weaknesses of the two newcomers who are playing a leading role in the race toward AV commercialization.

|  | Waymo/Google | Tesla |
|---|---|---|
| Industry sector expertise | An Internet search company<br>Expertise in all facets of IT<br>Software and media company<br>Advertising a major source of revenue<br>Interest in leading future technologies | A battery company that makes electric cars<br>Vertically integrated "newcomer" OEM<br>Interest in solar and space technology |
| AV products and services | Interest in retrofitting OEM cars as AVs Interest in TaaS<br>Licensing AV OS and sensors | Build and sell autonomous electric AVs to consumers |
| Current major revenue sources | Advertising, licensing software and smartphones | Cars and energy products—batteries and solar (Powerpack, Powerwall, etc.) |
| Financial | Huge | Small revenue base so far |

|  | Waymo/Google | Tesla |
|---|---|---|
| strengths | Makes a lot of money | Not profitable in 2017 |
| Market value | Varies; $700 billion in 2017 | Varies a lot; $50 billion in 2017 |
| Target AV | Targeting SAE level 5, has on trial prototype OEM vehicles equipped with Waymo HW/SW | Builds full AV from the ground up. Has announced 2019 launch date for SAE level 4 |
| Strengths | Expertise in building operating systems. Excellent AI expertise. Huge financial clout. Sensor expertise. Connected AV experience | Experience in manufacturing cars. Has captured the attention of influential members of society. Led by a visionary leader. Has delivered on its vision |
| Weaknesses | Lack of experience in operating complex auto-manufacturing assembly lines. No auto sales, distribution and service network | Still not making a profit on cars. Long-term financial viability in question. Huge delivery bottlenecks on model 3 |

*Table 7.1: Waymo versus Tesla*

### 7.1.3 Apple

Apple has been in stealth mode so far with respect to its AV ambitions. While Apple has not publicly announced its plans, it's an open secret that Apple has been doing a fair bit of R&D on AVs. The only thing that CEO Tim Cook has said in a Bloomberg interview in June 2017 is that AI automation is an important priority for Apple. Apple has also obtained a license from the California DMV (Department of Motor Vehicles) to conduct AV testing there.

My view is that Apple will announce its entry into the AV business in due course—it is too big and too attractive an opportunity to leave for others. Apple doesn't need the endorsement of the market. The company will do it when it has an AV that can beat the competition. Apple can bring its unique design expertise and flavor to its brand of AVs—vertical integration of design, complete control on the manufacturing process, outstanding user interface and, above all, safety of the passengers. It even has the cash to buy one of the OEMs, if it so chooses. Stay tuned.

## 7.2 Established OEMs

In this category, we have the traditional original equipment manufacturers like Ford, GM, Daimler, etc.

## 7.2.1 Ford Motor Company

Ford is the second-largest automaker (behind General Motors) in the United States and the fifth-largest in the world (behind Toyota, Volkswagen, Hyundai-Kia and General Motors) based on 2016 vehicle-production statistics. From a historical perspective, Ford is the company that revolutionized manufacturing by inventing the assembly line in the early 1900s. Henry Ford's name is unanimous with the Model T and other auto innovations from that era. In 2016, Ford announced its intent to produce Ford Fusion–based AVs in 2021 for commercial use in ride-hailing service Lyft. Interestingly, GM is a major ($500-million) investor in Lyft.

*Figure 7.2: Ford autonomous car (Source: Ford)*

### Ford's AV Vision and Strategy

Ford plans to operate a future version of a Ford Fusion AV without a steering wheel, gas pedal or brake pedal within geo-fenced areas as part of a ride-sharing or ride-hailing experience. Ford's President of Global Operations Jim Farley announced in an interview in December 2017 on Medium portal that it will unveil a new car design with a smaller LiDAR mounted on top of the car in 2019. Ford is also working on using its AVs for package delivery fleet applications (possibly UPS or Fedex).

### Ford's AV Alliances

Ford has partnered with or invested in four different technology companies. It acquired SAIPS, an Israeli company focusing on machine learning and computer vision; it is partnering exclusively with Nirenberg Neuroscience to bring more "human-like intelligence" to machine learning components of driverless-car systems; it is working with Civil Maps, a start-up based in New York that specializes in 3D high-resolution mapping technology; and, finally, Ford is working with Velodyne. Ford also indicated that it will double its workforce in Silicon Valley, and is also working with universities (Stanford, MIT, Michigan and Aachen in Germany). And in February 2016, Ford invested $1 billion (to be spread out over five years) in Argo AI, an AI start-up based in Pittsburgh, to aid its autonomous car efforts.

Also in 2016, Ford signed an agreement with QNX (the Blackberry subsidiary based in Ottawa, Canada) for its expertise in building infotainment OS and autonomous car software. As part of this agreement, BlackBerry will dedicate a team to work with Ford on expanding the use of BlackBerry's QNX Neutrino Operating System, Certicom security technology, QNX hypervisor and QNX audio processing software. Informal reports indicate that the total team will consist of 400 members from the two companies.

**Ford's Timeline**
In a 2016 press release, Ford announced that it plans to introduce a fully autonomous car in 2021.

### 7.2.2 General Motors

General Motors is the largest North American auto OEM and sold 9.9 million cars in 2016. GM has announced its intent to enter AV market through Lyft in which it invested $500 million in a billion-dollar financing round in 2016. GM negotiated a seat on Lyft's board as well. Surprisingly, Ford also entered into an agreement with Lyft to use Fusion hybrid AVs in Lyft's fleet, so apparently GM's relationship with Lyft is not exclusive. Waymo/Google has invested in both Uber and Lyft, though it now has a sour relationship with Uber because Uber hired several of Google's key technical experts. Strange are partnerships and competitive scenarios in the AV marketplace. To make matters more interesting, Alphabet has invested $1 billion in Lyft, suggesting that Lyft may use Waymo's OS software instead of the one GM is presumably working on through its acquisition of Cruise Automation.

GM has been doing trials with 40 Chevrolet Bolts in Detroit, Scottsdale and San Francisco, and has an extensive trial program planned for 2019.

**GM's AV Vision and Strategy**
GM's head of planning and strategy, Mike Abelson, says that GM's AV vision is in "on-demand autonomous ride-sharing." He says that some people will buy private autonomous cars but that most early adopters will start with autonomous ride-sharing, using different cars for different occasions. According to GM, most of the autonomous vehicles will be in urban congested areas supplementing public transit, and many autonomous cars will be electric. Once they gain enough trust in the new technology, consumers may invest in the private AV.

GM's initial AV strategy is based on getting its new Bolt electric cars into Lyft's fleet for driverless operation. It does not want to sell Bolt AVs to private consumers in the first phase. Reading between the lines of press releases and interpreting what the GM executives say and don't say publicly, it seems clear that GM does not think that all the regulatory, insurance and other issues will be resolved in the near term.

Abelson sees GM's experience with OnStar in the cellular cloud as a great advantage in terms of customer-interface protocols. While OnStar experience will help, the connected AV is a different beast and poses many other OTA communication requirements that GM has to learn and gain expertise in.

According to the latest research report (Q1 2018) from Navigant, a consulting company in the United States, GM has leapfrogged others in becoming a leader in the AV race. Waymo was second.

**GM's Alliances**

To ensure a strong entry into the AV market, GM has either invested in or partnered with a number of companies and ventures.

- GM spent $1 billion to buy Cruise Automation, a San Francisco technology company. Cruise is working on an autonomous version of the Chevrolet Bolt, which will be used in Lyft's fleet under the GM brand.
- GM made a $500-million investment in ride-sharing company Lyft. Lyft is an essential part of GM's Maven car-sharing service, which was operating out of Ann Arbor, New York City and Chicago as of 2018. With this approach it will serve different incarnations of TaaS.
- GM's Cruise Automation is also trying to do a deal with Uber.
- GM has set up a large technology R&D center in Silicon Valley and is looking to staff it with over a thousand highly technical employees.

*Figure 7.3: A GM Cruise AV (Source: GM)*

**GM's Timeline**

As for when GM will launch its AV service, the company is not giving a firm timeline and says that a lot will depend on regulators. However, according to one press rumor (which hasn't been confirmed or denied), GM will have thousands of AVs in trial in Lyft's fleet (semi-autonomous with drivers, I assume) in 2018.

### 7.2.3 Daimler Mercedes

Daimler is considered the inventor of automobiles, having built the first car 130 years ago, before Ford did it in the United States. Daimler is also perhaps the only established OEM that has been involved in academic-cum-industry AV-related R&D in Europe since the late 1980s. Daimler was an integral industrial partner in Europe's $900-million Eureka PROMTHEUS project—much earlier than any other company. The company has a wealth of experience in building luxury cars with advanced intelligent-drive safety features, which are used in semi-autonomous cars today but are on their way to fully autonomous versions.

*Figure 7.4: Daimler's F 105 "Luxury in Motion" concept car (Source: Daimler website)*

Daimler has recently demonstrated its S 500 Intelligent Drive, the F 105 "Luxury in Motion" concept research car (shown in Figure 7.4) and its Future Truck 2025 model equipped with intelligent "Highway Pilot" software (shown in Figure 7.5), all branded under Mercedes-Benz.

*Figure 7.5: Daimler's Future Truck 2025 (Source: Daimler)*

**Daimler's AV Vision and Strategy**
Daimler's current vision for autonomous cars is incremental and gradual. It will introduce semi-autonomous cars first, but has not publicized any timeline so far. Daimler doesn't feel that the industry is mature enough yet to turn the entire control and responsibility over to the Daimler "Drive Pilot" AI software.

Daimler's AV strategy revolves around the acronym CASE—Connected, Autonomous, Shared and Electric. Daimler's initial offering in the AV marketplace is relatively pragmatic and is based on "robo-taxis" in Uber's fleet. Daimler feels that the shared AV business model will be easier to implement than selling AVs to private luxury-car consumers. The company believes that if consumers gain trust in AVs after ride-sharing, they might then considering buying luxury AVs for themselves if their car usage is high enough to warrant dedicated ownership.

**Daimler's Alliances**
- Daimler signed an agreement with tier-1 industry supplier Bosch in 2017 to develop the necessary hardware and software to make its flagship luxury cars fully autonomous. The focus of the partnership will be on the software and algorithms required to make advanced driving systems safe and predictable.
- In early 2017, Daimler also announced an agreement with Uber[23], saying that it will use its "robo-taxis" in Uber network for its TaaS offering.

Daimler will start with hybrid AVs and then switch to all-electric car as the range of electric cars with one charge increases.

**Daimler Strategy**
Daimler's AV strategy revolves around acronym CASE—C for connected, A for "Autonomous", S for Shared and E for electric. Daimler's initial offering in AV marketplace is relatively pragmatic and is based on "Robo Taxis" in Uber fleet. Daimler feels that shared AV business model would be easier to implement than selling the AV's to private luxury car consumers. If consumers gain trust in AVs after ride-sharing, they might buy luxury AV if their car usage is high enough to warrant dedicated ownership.

### 7.2.4 Established OEMs—BMW

BMW is a German luxury-car OEM competing with Daimler and Lexus. It has investigated whether it should pursue building in-house expertise in AVs or buy the required system-integration expertise through partnerships with tier-1 AV component suppliers. After a lot of investigation, BMW management chose the latter in July 2016. According to Reuters, Klaus Büttner, BMW's vice president of AV projects, said that "many of us are swimming in the same sludge. Everybody is investing billions. Our view was that it makes sense to club together to develop some core systems as a platform."

BMW decided to seek the help of Intel's Mobileye to develop its AV offering. Meanwhile, it is showing up at trade shows such as the 2017 Consumer Electronics Show (CES) with demonstrations to keep the public's interest and let its customer base know that BMW is in the AV game (see Figure 7.6).

*Figure 7.6: BMW AV (Source: BMW at CES 2017)*

**BMW's AV Vision and Strategy**

BMW does not have a unique vision of its own—its strategy is to let the AV market decide where it is going. BMW will catch up with the help of external AV component suppliers.

**BMW's Alliances**

BMW has teamed with Intel/Mobileye to build its AV offering.

**BMW's Timeline**

BMW is hoping to offer an AV solution by 2021.

### 7.2.5 Volkswagen Audi

According to *Forbes* data, Volkswagen became the largest auto company in the world in 2016, beating long-time leader Toyota by about 100,000 cars (VW sold 10.3 million; Toyota, 10.2 million). Audi is the luxury-car division of the VW family.

In early 2017, Volkswagen CEO Matthias Müller indicated that VW group plans to launch a number of fully autonomous, SAE level-5 electric cars, vans and trucks by 2021. VW group has showcased two concept cars at recent trade shows: SEDRIC (Self-DRIving Car; see Figure 7.7), a cheaper utility car, and the Audi Aicon, a luxury car (see Figure 7.8). "Our team is already working on ideas for an entire SEDRIC family," Müller said, as reported on VW website, "This will range from fully autonomous vehicles for the city, for luxurious long-range mobility, through self-driving delivery vans and heavy commercial trucks."

*Figure 7.7: VW's SEDRIC AV Concept Car (Source: VW)*

*Figure 7.8: The Audi Aicon AV Concept Car (Source: VW)*

Both of these cars will be without a cockpit, steering wheels or pedals. While SEDRIC was conceived as a shared vehicle, it could be configured to serve an individual owner, too, VW has said.[24]

> According to VW's head of digitization, Johann Jungwirth, the company plans to have fleets of fully autonomous ride-sharing vehicles running in two to five cities around the world by 2021, running to all addresses in those cities via the VW-affiliated Moia and Gett ride-hailing brands.
>
> The Aicon, which debuted in Frankfurt in 2017, is nearly 18 feet long and seven feet wide. The company's vision is make riders feel like they're traveling in a first-class airline cabin. When passengers get out of the car at night, a minidrone accompanies them to light their way. The vehicle does not have long-range headlights since its laser and radar sensors can see in the dark, it can drive up to 500 miles on just one charge, and it can recharge wirelessly via induction.
>
> The Audi website covers some other cool features of this luxury AV: "The Audi Aicon supports its surroundings intelligently and uses animations on its display surfaces to warn pedestrians or cyclists of dangerous situations. Driving modes such as platooning, urban driving or driving at a walking pace can be visualized. Horizontal stripes of light move from the bottom up when the car accelerates and

in the opposite direction during braking. Their speed increases or decreases in sync with that of the car."

## VW and Audi's AV Vision and Strategy

VW and Audi intend to compete in all sectors of the AV market (ride-sharing fleets, individual owners and luxury owners) as it develops. Initially VW is going after the ride-sharing market and replacement of smaller public transportation market like buses.

## VW and Audi's Alliances

- Nvidia and Audi plan to bring a fully autonomous vehicle to the market in 2020–2021.
- VW and Audi are teaming up with mapping firm HERE, along with Daimler and BMW; HERE is a German company, which originally resulted from the merger of Navteq and Nokia Maps.
- Stanford University's CARS (Center of Automotive Research at Stanford) is home to the Volkswagen Automotive Innovation Lab (VAIL).

## VW and Audi's Timeline

As mentioned above, VW/Audi have announced that VW's SEDRIC and Audi's Aicon cars/fleet vans will be running in two to five cities globally in 2021. It appears that these are target dates.

In my view, this will be a challenge. The VW/Audi group has announced an ambitious AV vision in three or four vehicle segments. While they have the R&D resources, as the largest or the second-largest auto OEM in the world, they have not shown enough evidence so far to support this vision and timeline. They'll have a lot of work ahead of them if they want to convince the market that these timelines are realistic.

### 7.2.6 Volvo (*Source – Volvo, AutoNews and wired.com Eric Adams 2017*)

Volvo is a Swedish auto company by pedigree but was bought by Ford Motor Co. in the recent past. Then Ford sold it to China's Geely Holding Group in 2010. Volvo has revitalized itself recently. Global sales in 2016 stood at 534,000 units with plans to grow to 800,000 by 2020.

## Volvo's AV Vision and Strategy

Volvo cars are known for their sturdiness and safety though historically less for their aesthetic appeal. So it's no surprise that Volvo's public declarations about entering the AV market center around its vision to eliminate deaths in its cars by 2020. Volvo's CEO also announced in 2017 that Volvo would stop designing and producing ICE cars entirely after 2019, implying that all new Volvos will be electric or hybrid.

Volvo intends to start with a semi-autonomous XC60 that detects obstacles and brakes in emergencies using ADAS features and an early version of its own autopilot software. It will actively steer itself back to a safe lane if the driver risks a head-on collision by turning left, and it will help the driver weave around pedestrians, animals and other vehicles. It should be said that Volvo believes in making the human driver ultimately responsible for taking over in emergencies when the Volvo autopilot gets into trouble.

**Volvo's Drive Me Program** *(Source Vilhelm Carlström of Nordic Business insider)*
In order to get user feedback, in January 2017 Volvo announced a unique consumer research program called "Drive Me" that would let real users drive autonomous cars in Sweden. Under this program, Volvo was to give XC90s equipped with autonomous driving hardware and supporting software technology to 100 ordinary drivers.

*Figure 7.9: Volvo Drive Me XC90 Trials in Sweden (Source: Volvo)*

The Drive Me program was expected to examine in a holistic fashion not just technological issues but also infrastructural, societal and user issues. It was expected to answer questions like:

- What positive effects can autonomous driving have on the environment, safety and traffic flow?
- What requirements do AVs pose on infrastructure?
- Which traffic situations are suitable for AVs?
- What confidence do consumers have in self-driving vehicles?
- How can other drivers interact with AVs?

In December 2017, after a partial rollout of the program, Volvo announced that due to safety concerns the program would be delayed till 2021.

**My View on Volvo's Strategy**
I like Volvo's emphasis on electric cars as the base for their AVs, and also its emphasis on perfecting ADAS safety features before shooting for the distant goal of full autonomy. In the first phase of Volvo's AV introduction, it plans to leave the human driver in control. The company is not claiming to be the leader in the industry's race toward full autonomy but wants to stay abreast of technology innovations and keep pace with consumer

acceptance and adoption—to be led by consumers rather than try to lead them. Waiting until all the handover problems are resolved in level 4s and the last word is written on level 5s is a very realistic approach. Volvo's partnership with Uber is also significant as it includes the understanding that Uber won't take human drivers out of Volvo AVs without Volvo's permission. Volvo's Drive Me program is creditworthy. While it's somewhat similar to Waymo's program in Phoenix, Arizona, Volvo is giving users more flexibility than Google is in its trials. Google's approach is more paternalistic, while Volvo intends to modify its AI software by incorporating input from users. Volvo will get better insight into consumer trust issues through this program.

**Volvo's Alliances**
- Volvo announced an alliance with Uber in 2016
- The company has partnered with Swedish supplier Autoliv for driver-assistance technology (essentially ADAS)
- Volvo has a program with Berkeley University, with Penn State University as a sub-contractor
- It has also partnered with Nvidia for incorporating DrivePx supercomputer in its cars.

**Volvo's Timeline**
Volvo expects to introduce SAE level-4 AVs by 2021.

### 7.2.7 Toyota *Source (Toyota website)*

Toyota was the largest auto OEM in the world for several years, but depending on whose figures you use, Volkswagen and Toyota might have tied for the "largest OEM" title in 2016 and 2017.

For several years, Toyota opted to be a bystander in the AV race. The company didn't feel that the fully autonomous car movement would take off. It treated Google as an upstart that could not rock the auto establishment. Instead, senior management decided that Toyota should concentrate on making cars safer through ADAS features only. However, in a significant turnaround, Toyota acknowledged in 2015 that it must prepare for the inevitable introduction of autonomous vehicles in the next decade. The company allocated nearly a billion dollars to this effort—a similar sized investment as that of GM and Ford.

**Toyota's AV Vision and Strategy**
Toyota has published a document describing its approach toward AV technology, in which it describes two concepts for the human–car relationship—the chauffeur concept and the guardian concept—as follows:

> ***Toyota Chauffeur Capability*** *is a measure of the degree to which the vehicle takes primary responsibility for driving, relieving the human driver of some or all driving tasks. If the Chauffeur capability is low, the human may be responsible for monitoring the environment and acting as a fallback (similar to SAE Level Two). If the capability is a bit*

higher, the human driver may be responsible only for acting as a fallback (similar to SAE Level Three). If the Chauffeur capability is high, the human driver may have no responsibility at all (like SAE Levels Four and Five).

**Toyota Guardian Capability** *is a measure of how much the automated driving system helps to protect occupants while the human is driving—both from mistakes or other errors by the driver and from external factors on the road such as vehicles, obstructions, or traffic conflicts. The higher the Guardian capability, the greater the number and types of crashes it can help protect against. For example, at a modest level of Guardian capability, systems like Lane Departure Alert (LDA) and Automatic Emergency Braking (AEB) can help prevent some crashes. At the highest level, Guardian capability would help ensure a vehicle driven by a human being would never cause a crash, regardless of any error made by the human driver, and steer a vehicle to avoid many crashes that would otherwise be caused by other vehicles or external factors.*

*Although Chauffeur and Guardian capabilities reflect distinct concepts for automated driving, their development builds on similar perception, prediction, and planning technology. Indeed, the hardware and software required for Guardian capability serves as the backbone for Chauffeur capability.*[25]

Figure 7.10 : A Toyota autonomous car (Source: Toyota)

### Toyota's Research Initiatives and Alliances

- Toyota has established the Toyota Research Institute (TRI), headed by Gill Pratt, with a billion-dollar investment. It has set up offices in Michigan, Silicon Valley and Cambridge, Massachusetts. TRI is focused on AVs and robotics.
- Toyota has also partnered with the University of Stanford (in a $25 million research cooperation), MIT and the University of Michigan.
- Toyota AI Ventures (part of TRI) has made five investments in AI-related companies, including Nauto and Intuition Robotics (an Israeli AI company), and expects to invest in more worldwide.

At Stanford University, Toyota is engaged in several interesting AV projects. The company is testing cameras that will watch the driver inside a car, and AI apps that engage in conversation with the driver and keep him/her awake. If drivers get drowsy, the automated safety systems will bring the car to a safe stop. Toyota's TRI is testing "guardian angel" technology that would take evasive steps in the case where a human driver is unable to handle imminent trouble. Toyota is also working on an AI system called Yui that, among other things, helps keep the driver mentally alert by engaging him/her in routine tasks like talking or turning on the radio. Yui can even measure human emotion through face recognition, and smoothly transfer control to/from the human driver.

### Toyota's Timeline

Toyota has indicated only a tentative timeline for its AV implementation. It says it will offer Highway Teammate in 2020 and Urban Teammate later in the 2020s—this could mean as far out as 2028.

### My View on Toyota's AV Strategy

Toyota is certainly not a leader in the AV race. It recognizes that there are still many hurdles to leap before we achieve fully autonomous driving, and suggests that the industry is nowhere close to SAE level-5 capability. Nonetheless, it has allocated enough R&D dollars to keep itself abreast of the advances and jump in if the AV market takes off.

### 7.2.8 FCA (Fiat Chrysler Auto)

FCA group is another large auto OEM that combines the manufacturing and sales distribution networks of Dodge Chrysler in North America and Fiat in Europe. Famous auto brands under FCA group are Chrysler, Dodge, Fiat, Jeep, Alpha Romeo, Mopar and several others.

### FCA's AV Vision and Strategy

Unlike Ford, GM, Daimler and VW Audi, FCA has been very passive about autonomous car technology. The company released a very non-committal press release downplaying the progress of AV front-runners and suggesting that fully autonomous cars are further into the future than others suggest. The news release emphasized some of the ADAS features on FCA cars, such as a forward-collision warning system and Automatic Emergency Breaking (AEB).

However, FCA is lucky in that it has a relationship with Waymo. While Waymo tried to strike a deal with Ford initially, it appears that it found FCA to be a more willing partner—perhaps because FCA had no major initiatives and ambitions of its own. Even though the relationship is not exclusive, Waymo is doing trials with 100 Chrysler Pacifica vans in Phoenix, Arizona. Some reports indicate that the partnership may increase the number of FCA vans under Waymo trials to 500. This increase in trial vehicles would give Waymo the "big data" it needs for its deep learning–based "Waymo-pilot" AI software, which would allow it to reach the SAE level-5 AV goal-post sooner than others. If FCA can take advantage of this relationship and equip its people-mover vans with the sensors that Waymo recommends, it might turn out to be a winning proposition for both FCA and Waymo—most analysts believe that in the first phase of AV adoption, fleet applications will carry the day before private AV customers start getting in line.

*Figure 7.11: FCA's Chrysler Pacifica people-mover (Source: FCA)*

The FCA people-mover is expected to be an electric van with a 250-mile range and an interior that's significantly equipped with technology designed to appeal to younger generations. I can speculate that it will be suitable for TaaS fleet applications.

### FCA's Alliances
FCA's relationship with Alphabet/Waymo is working well.

### FCA Timeline
FCA has provided a tentative date of 2020, and I understand that their target is SAE level 4.

### 7.2.9 Renault-Nissan-Mitsubishi Alliance *(Source – Nissan)[26]*

The Renault-Nissan-Mitsubishi Alliance is now the largest (note – statistics dissemination companies treat this alliance as three OEMs) seller of automobiles in the world. The three OEMs operate as three separate but cooperating companies with a layer of common management, headed by Carlos Ghosn. Each company owns shares in the others, and they share common R&D in certain strategic areas—like AVs.

## Renault Nissan Mitsubishi (RNM) Group's AV Vision and Strategy

Led by Nissan, the group has been pursuing a more aggressive approach to the AV race than Japanese competitors like Toyota. The company has been involved in several trials of ADAS features under the "Nissan Safety Shield" brand. Nissan is now testing a prototype vehicle that can perform all typical driving maneuvers including merging and passing, and that may contribute significantly to reducing accidents caused by human error and inattention. Nissan doesn't agree with Google's approach of eliminating the human driver completely this early in AV development. Instead, Nissan's program (called Nissan Intelligent Mobility) is focused on a future in which cars and drivers are partners in a more confident and more connected driving environment.

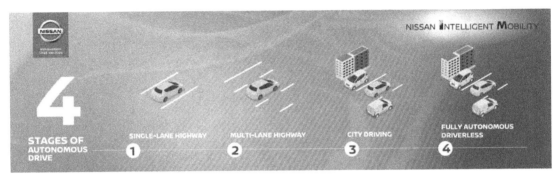

*Figure 7.12: Nissan's vision of four stages of autonomous driving (Source: Nissan)*

Using technology originally developed by NASA, according to Ghosn's announcement at CES 2017, Nissan is working on a Seamless Autonomous Mobility system (SAM), which combines "in-vehicle Artificial Intelligence (AI) with human support to help an autonomous vehicle make decisions in unpredictable situations."

## RNM Group's Alliances

Also at CES 2017, Ghosn mentioned the following partnerships or alliances that are helping to inform the group's AV efforts:

- The group is partnering with the US Department of Defense and several top universities, including MIT, Stanford, Oxford, and the University of Tokyo.
- They are working with Microsoft in adapting the Cortana human speech recognition capabilities for an HMI. According to Ghosn, Cortana will enable a vehicle to adapt to personalized driver settings and recognize different driver preferences even in a shared vehicle.
- Nissan is also collaborating with Bose to develop technology that increases drivers' situational awareness. Bose's "Aware" technology helps drivers react to sounds such as safety prompts, navigation signals, system alerts, texts and phone calls, with speakers integrated into the headrest that deliver relevant sounds to the relevant ear

(the left turn signal is heard in the left ear, etc.) and deliver phone-call audio to the seat of the passenger receiving the call.

**RNM Group's Timeline**

FCA CEO Ghosn had announced that the group will have a level-4 AV by 2020.

## 7.3 A New Breed of AV OEMs

Several new start-ups with the intent of assembling and marketing fully autonomous vehicles in the luxury-racecar category have been funded over the last several years.

*Figure 7.13: NIO concept car (Source: NIO.com)*

### 7.3.1 NIO

NIO is a Chinese-backed start-up formerly known as NextEV. At a 2017 event in Austin, Texas, the company's US CEO Padmasree Warrior announced plans to bring an autonomous electric car to the United States by 2020.

NIO has been building capabilities around the world, including an autonomous-driving office in San Jose, a design center in Germany and a motorsport center in the United Kingdom, and is partnering with Mobileye, Nvidia and NXP Semiconductors.

NIO's self-driving car will have level-4 autonomy in a constrained environment, according to *Forbes*, and "will first be introduced to the US market followed by Europe. ... The vehicle will have a steering wheel and pedal so its human occupants can take over driving if they want. NIO didn't announce a price, although Warrior says the company is targeting the premium segment."

### 7.3.2 Baidu

Asian search engine giant Baidu and auto manufacturer Changan have joined the AV race. In 2016, Changan cars made a 1,200-mile trip from the southwestern city of Chongqing to Beijing, marking the nation's first long-distance test of this technology. And now, Baidu has unveiled plans of its own to start mass producing autonomous cars by 2021.[27]

## 7.4 Tier-1 AV Component Suppliers.

In this category, we list established tier-1 component suppliers who are integral pat of the AV landscape, generally in support of the OEMs.

### 7.4.1 Bosch

Bosch is a large Tier-1 auto component supplier and is now building an important AV component—an AI supercomputer using Nvidia Drive PX chips with Xavier architecture. Xavier can process up to 30 trillion deep-learning operations a second while drawing only 30 watts of power.

At the beginning of 2017 Bosch also announced that it will invest $400 million by 2021 in expanding its expertise in the area. This includes establishing a Center for Artificial Intelligence across sites in India (Bengaluru), the United States (Palo Alto) and Germany (Renningen).

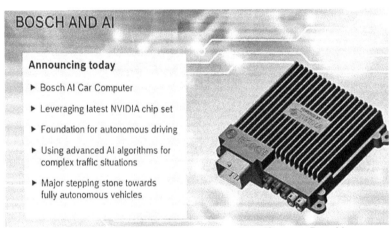

*Figure 7.14: The Bosch AI supercomputer (Source: Bosch)*

### 7.4.2 Delphi/Aptiv

Delphi is one of the largest automotive parts manufacturers headquartered in the United Kingdom (it used to be the "parts" portion of GM, and it has approximately $17 billion revenue and 150,000 employees). The company was recently split into two separate companies—the traditional parts company and a new company providing AV-focused components and services to auto OEMs. The latter is known as Aptiv and will focus on the parts of the Delphi business that deal with integrating computer controls and ADAS components into vehicles. Delphi paid $400 million in 2017 to acquire NuTonomy—an MIT spin-off technology start-up that makes software for building self-driving cars and autonomous mobile robots, which had a staff of 70–100 engineers at the time it was acquired. This acquisition will become a part of Aptiv, presumably. NuTonomy is running

AV taxi trials in Boston and Singapore. Delphi/Aptiv has relationships with other companies such as Intel/Mobileye and sensor manufacturers.

### 7.4.3 Continental AG  *(Source – Continental)*

Continental AG is a major German automotive manufacturer that specializes in tires, brake systems, interior electronics, automotive safety, powertrain and chassis. It had revenue in the $50-billion range in 2016 with 220,000 employees.

Continental has entered the AV race and is developing solutions for driverless vehicles in urban areas, having introduced the CUbE (Continental Urban mobility Experience) demo vehicle as a robo-taxi. The idea behind the electric CUbE is to provide a "feel-good cocoon" with "interiors supporting individualization but with functional integration in surfaces." Continental envisions that CUbE-like vehicles will contribute to modern mobility, solving congestion, accidents, contaminated air and parking problems in cities. Continental plans to test and research driverless transportation at its Frankfurt location in the future.

*Figure 7.15: Continental's CUbE for AV fleet applications (Source: Continental)*

In 2017, Continental announced autonomous car collaboration with Baidu.

### 7.4.4 Magna

Magna is a global automotive supplier with Canadian roots. It has 327 manufacturing operations and 100 product development, engineering and sales centers in 29 countries, and over 161,000 employees.

**Magna Steyr**[28] (a Magna subsidiary) is working on "future functional and electric architectures" applicable to AVs, working with industrial and academic partners. The company's website lists the following development areas:

- *"Vehicle integration of future Advanced Driver Assistance Systems (ADAS): concepts, virtual and real world testing strategies and maneuver catalogues on system and vehicle level*

- Concepts and demands for integration and validation of future semi- and fully autonomous driving systems
- Future data handling and partitioning concepts (CAN, Flexray, Ethernet, etc.) and external connectivity via LTE, 5G, WiFi, Car2Car-communication, etc.
- Optimization and integration of new HMI concepts to reduce driver distraction and getting driver back in the loop after automated driving situations
- Simulation of electrical energy management including 48V and mechatronic systems for enhancing development quality and reducing development time"

## 7.5 Computer-centric Tier-2 AV Component Players

The following automotive computer chip and ADAS component players are playing increasingly important roles in computer-centric components for AVs.

### 7.5.1 Intel/Mobileye *(Source: Mobileye website)*

In 2016, Intel, the Silicon Valley computer chip giant, surprised the investment community by purchasing Mobileye, an Israeli ADAS company founded in 1999 (390 million revenue 450 employees) for 15.3 billion dollars. It was a major vote of confidence from Silicon Valley for Mobileye's role in developing computer vision–based ADAS solutions for autonomous cars. Mobileye equipment had been installed in over 15 million vehicles by 2016.

Mobileye has invested over 15 years in R&D to gain insight into ADAS and fully autonomous vehicles meeting SAE levels 4 and 5. According to the company's website, Mobileye is a major player in "the development of monocular vision-based advanced collision avoidance systems, providing system-on-chip [SoC] and computer-vision algorithms to run Driver Assistance Systems (DAS) functions such as Lane Departure Warning, Vehicle Detection for radar-vision fusion, Forward Collision Warning, Headway Monitoring, Pedestrian Detection, Intelligent High Beam Control, Traffic Sign Recognition, vision-only Adaptive Cruise Control, and more."

The combination of Intel's leadership in silicon chip design and manufacturing infrastructure with Mobileye's proven expertise in ADAS and AV hardware/software will make them a formidable player in the AV landscape.

### 7.5.2 Texas Instruments

TI is an important silicon supplier and has started designing and manufacturing AV-specific silicon components. I have described TI contribution to AV industry in chapter 5.

## 7.5.3 NXP

NXP is the largest supplier of computer chips, embedded controllers and sensors to the automotive industry, with 14% market share. NXP is now owned by Qualcomm and has over 45,000 employees in more than 35 countries. It is getting deeply involved in the AV electronic component marketplace, especially in deep-learning platforms, directly in competition with Nvidia. Seeing a huge growing market, NXP is developing an autonomous vehicle framework BlueboxNXP at the basic silicon level i.e. low level chips – the same business NXP is in other areas. What NXP is developing is a general-purpose development platform that individual OEMs or tier 1 component suppliers can use to create custom solutions for their clients.

**My view of NXP**

I understand that the NXP framework is still conceptual and a "work in progress." I feel that NXP does demonstrate a good understanding of the requirements. The system-integration effort for AVs is *not* easy and must not be taken lightly for a mission-critical system like a level-5 autonomous vehicle.

### 7.5.4 Nvidia *(Source: Nvidia and Verge)*[29]

Apart from sensors that provide information required about the surrounding environment and any obstructions in a car's path, AVs need a supercomputer that is fast enough and smart enough to absorb that information and make decisions that are fed into the actuators that physically move, steer and brake the vehicle. This supercomputer and its deep-learning AI software (often called the brain of the system) is one of the most important elements of the self-driving car puzzle, and it's where Nvidia plays.

In 2015, Nvidia announced its Drive PX supercomputer platform to support ADAS features in luxury cars. As a result of the Drive PX's success, Nvidia was able to establish partnerships with over 200 companies (including major OEMs)—Toyota, Audi, Tesla, Mercedes, Volvo, Baidu, Bosch, ZF, AutoLiv, BMW and Luxgen (Taiwan). In 2016, Nvidia upgraded the Drive PX to Drive PX2, adding more horsepower, which meant —meaning more AI operations per second.

While Drive PX has met the challenge for handling ADAS features, the AV industry architects realized that the requirements for fully autonomous cars meeting SAE level 4 and 5 will be much greater than those of ADAS featured cars (level 2 and 3). To meet these increased computing requirements (hardware power and AI data size cum software complexity), Nvidia made a spectacular announcement in October 2017 of Drive PX *Pegasus* that will be ten times faster than its predecessor. Pegasus is capable of performing 320 trillion deep-learning operations per second, a tenfold increase over the Drive PX2. Nvidia has downsized the real-estate and power requirements of the supercomputer as well—it is the size of a car license plate. This used to be a major headache facing computer architects working on AVs.

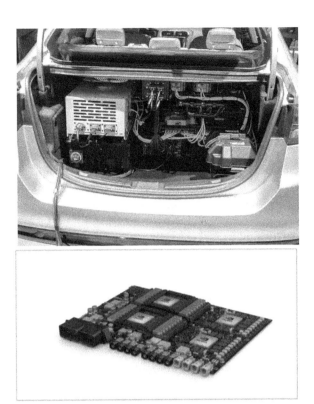

*Figure 7.16—Nvidia Drive PX Pegasus Supercomputer in the trunk of Ford Fusion (Photo courtesy Nvidia website)*

Pegasus achieves this level of performance by including four processors, including two Xavier SoCs featuring eight custom CPU cores and two next-generation discrete GPUs. Pegasus will be available to Nvidia automotive partners during the second half of 2018, including the dozens of companies trying to bring self-driving robo-taxis to market as quickly as possible.

Of course, Nvidia faces tough competition from Intel which has been building supercomputers using Xeon processors and field programmable gate arrays, and Waymo, which has been working on custom Tensor Processing units instead of GPUs. Yet, Nvidia has an early fast-mover advantage and if the early trials of its partners/clients prove that it can deliver on its potential, it will retain its leadership role.

## 7.6 QNX

I have discussed the role of QNX and its OS adequately in chapter 6.

## 7.7 Specialty AV Players

There is a whole slew of players who have gotten or are getting into the AV market. In fact, there may be thousands of players in this landscape. I won't elaborate on these players but I'll just mention that the AV connectivity players and app developers will play an important part in the AV market.

Figure 13.1 in chapter 13, designed by Liz Jensen of VentureBeat in Silicon Valley, shows the incredible width and breadth of the AV landscape. She calls it "Connected Car" landscape – still another perspective and another name for AV Plus.

**Summary**
*Here, I have briefly described the AV landscape—including both established OEMs and upstart technology challengers. Some OEMs are actively and aggressively accepting the challenge, while others are taking a cautious approach. It does appear to me that the incremental strategy of established companies is more pragmatic considering the reluctance of current generations to let robo autopilots drive them to their destinations. Let's continue to innovate and introduce change at a rate that we can all absorb without too much stress.*

---

**Citations for External References**

[22] Inside Waymo's Secret Training of Self-driving Cars --
https://www.theatlantic.com/technology/archive/2017/08/inside-waymos-secret-testing-and-simulation-facilities/537648/

[23] Daimler Announces Agreement with Uber –
https://www.usatoday.com/story/money/cars/2017/02/02/uber-mercedes-benz-parent-partner-self-driving-cars/97406166/

[24] About VW strategy -- AutoNews
http://europe.autonews.com/article/20170917/COPY/309179994/volkswagen-readies-its-robo-fleets

[25] Toyota AV Strategy from Toyota website
http://d2ozruf16a8he.cloudfront.net/15/e430760e466468024f1f516b14c3eba0de22e7b9/PDFToyota_AutomatedVehicles_DL.pdf

[26] Nisan AV Story -- HTTPS://WWW.NISSANUSA.COM/EXPERIENCE-NISSAN/NEWS-AND-EVENTS/SELF-DRIVING-AUTONOMOUS-CAR.HTML

[27] Baidu Plans for AV – https://www.digitaltrends.com/cars/china-self-driving-cars-baidu-changan/

[28] Magna Stehr website – http://www.magna.com

[29] Nvidia's Drive Pegasus Px Platform – https://www.theverge.com/2017/10/10/16449416/nvidia-pegasus-self-driving-car-ai-robotaxi

# Chapter 8

# TaaS—Concept and Players

*AV players are hoping to transform our traditional method of transportation based on exclusive-use and privately owned cars to one based on a shared utility that the industry calls TaaS. They want to make it economical and convenient to call in an AV at a moment's notice so that you might never feel the need for your own car. Call it, hop in, drive away to your destination. It's Uber or Lyft ride-hailing without the Uber driver. TaaS will definitely wipe out the taxi industry. At what pace TaaS will dissuade us from buying cars in the future will depend on a number of factors. Another major motivation for TaaS is that it will allow consumers to try out autonomous driving without buying an AV. Fear of AVs will go down; trust will increase. Even if the current business model of TaaS proves to be flawed, the industry has lots of levers to manipulate the cost of TaaS. It'll be very interesting to watch this one.*

## 8.1 TaaS

Transportation as a Service, also called Mobility as a Service by some, is a concept that essentially means a shift from our current emphasis on personal ownership of cars to an emphasis on third party private or public transportation services, paying a single-use or ongoing charge for majority of our transportation needs. Under this type of arrangement, consumers may find that the cost of transportation from private and public transportation providers is lower than the cost of owning a car, not to mention that the service will be convenient and timely. At the same time, the obligation to keep one's personal car in a protected space and in road-worthy condition will no longer exist. Many families will no longer need garages, or certainly not two- or three-car versions. In a partial departure from the current car-ownership model for a modern urban family, many may find it more economical to switch to a hybrid model where they own just one car (instead of two) for urgent or long-distance travel needs, and use a combination of Uber-like services, on-demand public transit services run by the city and car-sharing services for the rest of their transportation needs.

This scenario becomes very attractive with the arrival of autonomous vehicles, where the cost of transportation services will be much lower than our current taxi or -like services. Today, the cost of a human taxi driver or Uber/Lyft driver is almost 50% of the total cost of the service. If we take the human driver out of the equation, we will reduce the cost of transportation service significantly. As autonomous vehicles become available at every intersection and we can easily just call them in with a click on our smartphones or a voice command to Alexa or Google home, the service will become more and more affordable and incredibly convenient.

The fundamental rationale behind TaaS is that personally owned vehicles are used only 4–5% of the time over a 24-hour period, sitting idle most of the time. From an economic point of view, that's a very low use of an expensive asset. As far as pride of ownership goes, that's mainly a phenomenon embedded in the minds of older generations; the desire to own a shiny car to show off to friends is seen much less in millennials and barely registers in Generation Z. For younger generations, personal car ownership is a burden. They'll be quite happy to treat cars as a utility that they can rent on a single, multiple or continuous-use basis. Just as individuals don't own power generators for their homes, they will no longer need to own their own cars.

Let us consider the following scenario:

- It's the year 2025, or maybe 2030—it remains to be seen when TaaS will move from a novelty service to a competitive service for the average household (see chapter 14 for my predictions). Just imagine—AVs have arrived. Uber, Lyft, Avis, Hertz, National and several other companies have been doing extensive trials and running free "Ride Me" or "Drive Me" promotional programs for several years to build your confidence and gain your trust in their TaaS services. They have a long list of testimonials and somebody in your group of friends has tried it. TaaS providers have been sending you 50% off coupons for several months. You're ready to try it.
- Your old (non-AV) car is ready to be retired after 10 years of faithful service.
- You've registered as a subscriber with your preferred service providers. They all have your payment authorization in their subscriber agreement.
- You have an app on your phone that allows you to call TaaS vehicles. Your friend uses voice commands with Alexa or Google (or whatever their new names are in this future world).
- Your app knows your personal preferences: you prefer Lyft for short daily commutes to the office and ZipCar for weekend trips to the zoo with the kids. If you're going to the airport, the app knows to direct you to the Tesloop site, which offers a shuttle service for a fixed price every 15 minutes (you select your pick-up time and Tesloop does the rest).
- With just a few clicks, you order a car, which comes on its own (without a human driver) from the neighborhood plaza or charging station that has parking spots for TaaS vehicles, where it's been waiting to be called. It arrives at your door within one to three minutes.
- The car sends you a message: "I am outside your house, ready to take you to the office or the zoo".
- If you asked for a shared AV, there may be other passengers in the vehicle.
- After the trip, the system asks you for 3-click or voice-activated feedback. (Competition is keen and TaaS providers want to hold on to customers by providing great service.)
- Your account is charged and you get a monthly invoice.

### The TaaS Business Model

Let's look at the economics of TaaS versus private ownership of a non-autonomous vehicle. Table 8.1 uses the following assumptions:

1. Purchase price: I've assumed that AVs will cost at least $10,000 more than non-AVs.
2. Financing cost: Assume money is borrowed at 5% and is paid off in five years.
3. Depreciation cost: I have assumed a straight-line method of depreciation.
4. Insurance cost: I've concluded that insurance costs will go down only slightly for non-AVs once AVs hit the landscape. As far as AVs are concerned, the industry has to allow for liability costs and property damage, whether or not these costs are hidden from consumers. I've assumed a 50% reduction for AVs.
5. Maintenance cost: While cars will be more reliable and service incidents will be less frequent, service personnel costs will increase significantly because of the complexity of repair and the skillset required.
6. Fuel costs include gas for ICE non-AVs and charging costs for electric cars and e-AVs. Gasoline costs $3.00 per US gallon and you get 25 miles per gallon.
7. The two types of vehicles will probably have similar registration costs.

|  | Current personal car-ownership model | Future TaaS model |
|---|---|---|
| Average cost of car | $40,000 for non-AV<br>$50,000 for AV | $60,000 retail price for AV (assume $10K for AV add-on) 20% off for TaaS providers |
| Average life of the car | 10 years | 5 years (because of extensive use) |
| Annual mileage | 15,000 miles | 100,000 miles |
| Depreciation cost (straight line) | 26.67 cents/mile assuming 10 year life and 15000 miles/annum | 10 cents @ 500K miles – life time<br>5 cents @ 1,000K miles – life time |
| Insurance costs—35-year-old driver, living five miles from work with a good driving record | $1,500/year = ~10 cents/mile (for non-AV) | Assume ~5 cents/mile (AVs are safer) |
| DMV Registration cost | $100/year = ~0.67 cents/mile | $100/year = approx. 0.13 cents/mile |
| Fuel cost/mile | ICE = 12-16 cents/mile (@ $3/gallon and 20-25 mpg)<br>Electric = 3-5 cents/mile | Electric = 3 - 5 cents/mile |
| Maintenance costs—annual | $400 for first 3 years, $800 after | $5,000 for AV |
| Maintenance costs—per mile | 2.5 cents; 5 cents thereafter | 5 to 10 cents per mile (500K/1000K) |
| Finance costs (borrowing @ 5% for 5 years) | $2,000 annually<br>Works out $13.3 cents/mile for 15000 miles average annual usage | $2,500 (50K@5%)<br>= 2.5 - 5 cents/mile |
| Additional—daily upkeep of the car | 0 (done by owner) | $10 per day = ~3 cents per mile |
| Remote support infrastructure, including 5G cloud charges | 0 | One cent for remote support plus one cent for cloud (per mile) |
| Total Costs (assume 100% markup over cost for TaaS model) | ICE: 69-70 cents/mile<br>Electric: 59-60 cents/mile | TaaS provider cost = 34 cents/mile<br>Assume 100% markup and add 10% for driving to client and parking location.<br>Customer cost = 74.8 cents/mile<br>(One Passenger), 37.4 cents per mile (two passengers) and 25 cents per mile (3 Passengers) |

Table 8.1: Comparison of economics of car ownership vs. TaaS (My assumptions) The reader may change some of the assumptions and compare costs.

Note: According to the Bureau of Transportation Statistics, the average cost to own and operate an automobile in 2016 is $8,858 per year, assuming 15,000 miles.

## Factors and Trends Fueling TaaS Model Adoption

There are a number of factors and trends that are driving the adoption of the TaaS model.

1. The oncoming reality of AVs might be the single most important factor making TaaS more affordable than owning a car. The economics of AVs without a human driver are significantly in favor of TaaS, which should result in reduced ownership of personal cars.
2. The popularity of current ride-sharing, ride-hailing and car-sharing services among millennials and Generation Z will lead to greater adoption of TaaS.
3. Convenience factor and less stress: AVs will allow TaaS users more time for other activities in a much lower-stress environment than when driving their own cars.
4. Integrated trip-planning applications are emerging that will allow end-to-end service with multiple sub-trip service providers, such as AV plus a public train-or-bus combination. A single unified payment system in which the customer gets one single invoice will make this very convenient.
5. The increasing popularity of environmentally clean electric AVs will persuade consumers to use TaaS.
6. There will be pressure on small public transit agencies to use on-demand TaaS services from private operators for low-traffic zones and low-traffic time periods instead of inefficient bus services with human drivers. Many of us have seen a 40-seater bus carrying just one or two passengers in the non-busy afternoon or late in the evening.

## My View on TaaS Replacing Private Ownership

While I agree with the general trend toward TaaS, I believe that this will happen slowly and gradually. Here is the scenario that I see unfolding:

1. The price charged for a transportation service (e.g., the per-mile charge) will likely be based more on competition and on what the customer will pay than on its cost. The *cost* of a service only indicates whether there's a business case for it. Therefore, TaaS adoption projections based purely on cost are highly flawed. Many TaaS reports (especially RethinkX's report, "Rethinking Transportation 2020–2030," by Tony Seba) are painting an extremely optimistic scenario for TaaS replacing private ownership.
2. I don't see TaaS providers offering their services at less than one dollar per mile. Taxi services are currently $2.50 to $3.00 per mile in North America.
3. Owners of private cars treat the purchase price and insurance as a sunk cost when they make trip decisions, only considering the variable cost elements involved (gasoline, wear and tear, etc.). Only when buying a new car do they consider all costs.

4. TaaS models and projections assume that all consumers will willingly accept TaaS rides even if they have to share the AV with other passengers. This may happen for certain kinds of trips (perhaps daily work-related ones) but not necessarily for all family trips.
5. TaaS providers will quite likely wipe out the traditional taxi industry within ten years from the introduction of services without a human driver.
6. Baby boomers will shy away from TaaS as a basic mode of transportation. TaaS will become popular with millennials to a small extent first and with Generation Z later to a large extent. Generation X (children of baby boomers) will behave in-between baby boomers and millennials.

## 8.2 Ride-Hailing and TaaS Providers: Uber

For the past five years, Uber has become synonymous with the ride-sharing or e-hailing services that have struck a huge blow to the traditional taxi industry in most large cities of the world. Uber was built on a business model that offered a less expensive and more convenient option to the public by allowing clients to order a ride in a private individual–owned car using a smartphone app. It was a great business model because Uber didn't have to own any vehicles of its own. Using modern computer technology and a solid marketing program, the company enlisted a huge number of private drivers with their own cars and, through them, provided a better and cheaper transportation service that undercut taxi-service prices.

Uber reported revenue of $3.4 billion in the first quarter of 2017. Although Uber hasn't been profitable so far, it has been expanding its network internationally and investing heavily in its future. In the third quarter of 2017, it was active in 84 countries and 737 cities. Its market valuation stood in the $70-billion range in 2017. This high valuation gives it tremendous room for raising funds to invest in its future projects.

As autonomous car innovation became a future reality, Uber saw an opportunity to reduce the cost of its ride-hailing service further by using autonomous vehicles. Since driver costs constitute almost 50% of taxi-service costs, Uber wanted to be the first on the block to take advantage of the TaaS business opportunity.

Uber is an aggressive company with ambitions not only in providing AV-based TaaS services on AVs from auto OEMs (it has entered into several agreements with OEMs) but in building its own brand of AVs. Uber set up ATG (Advanced Technology Group), which is investing heavily in R&D with the objective of creating proprietary hardware and software that it could potentially use to gain competitive advantage. ATG has set up centers in Pittsburgh, San Francisco and Toronto. In Pittsburgh, ATG hired more than 40 academic researchers from CMU, the premier center for AV research in the world. It also lured key talent away from Alphabet/Waymo and was involved in legal battles as a result—Alphabet alleged that Uber had stolen some of its proprietary technology and know-how. They subsequently settled the dispute.

Uber was one of the first companies to start running self-driving robo-taxi trials. Even though these trials, which started in Pittsburgh in 2016, currently have a human driver, the company's ultimate objective is to have driverless ride-hailing taxis. It should be noted that Uber's proposed business plan will change its asset ownership significantly – right now the company does not own vehicles. In future it will have to own these assets which will require huge capital.

### Uber Partnerships

- **Daimler Germany**: In June 2016, Uber announced an agreement with Daimler to use the Mercedes S-class cars in its ride-hailing network. The cars would be owned and operated by Daimler. This non-exclusive arrangement would give Uber's ride-hailing service a class of super-luxury cars. It would give Daimler some market experience about the popularity of luxury ride-hailing, in case it wanted to get into that business itself. Daimler is also testing autonomous S-class cars and trucks on its own.
- **Volvo:** Uber retrofitted Volvo's XC90 with its autonomous technology for trials in several cities.
- **Toyota:** Toyota made a strategic investment in Uber during 2016 and also announced collaboration with Uber's ride-hailing service.
- **Ford Fusion:** Uber has used the Ford Fusion as its base for trials in several cities.

## 8.3 Ride-Hailing and TaaS Providers: Lyft

Lyft is a TaaS provider based in San Francisco, California. Like Uber, it offers a mobile app that allows customers to hail rides. The company launched in 2012 and its services are now offered in over 300 cities in the United States as well as in Toronto (launched in December 2017). By 2017, Lyft was offering 18 million rides a month. The company was valued at $7.5 billion as of April 2017 and has raised a total of $2.61 billion in funding.

**Lyft Investor Partners:** A number of OEMs and other organizations have invested in Lyft as a hedge against Uber (which is leading the ride-hailing/TaaS marketplace).

- GM has invested $500 million in Lyft.
- Alphabet/Waymo has invested $one billion in Lyft.
- Ford has partnered with Lyft to use Ford Fusion AVs in the Lyft network.
- Lyft has also partnered with Jaguar, Drive.ai and NuTonomy.

Lyft is similar to Uber in many services it offers but it has also learned from Uber's mistakes (aggressive and unnecessary fights with regulators, not playing with taxi unions on equitable footing, not being sensitive to staff diversity complaints, not doing necessary police checks on its drivers, not complying with higher commercial insurance requirements that the public and regulators required, etc.).

Table 8.2 outlines the primary differences between the two major players in the TaaS market.

|   |   | Uber | Lyft |
|---|---|---|---|
|   | Key assets | Mobile app, dispatch app, relationships with drivers, and AV know-how | Relationships with Ford and GM (lukewarm now), better public perception, lean management, better focus, has learned from Uber's mistakes |
| 1. | Partnerships | Volvo, Ford Fusion, Mercedes | Ford, GM, Jaguar, Drive.ai, Waymo, NuTonomy |
| 2. | Coverage | 737 cities, 84 countries (2017) | 300 cities, US and Canada (2017) |
| 3. | Investment raised | $15 billion (2017) | $2.61 billion (mid-2017) |
| 4. | Valuation | $70 billion (2017) | $7.5 billion (2017) |
| 5. | Pricing | Competitive but fares lower than taxis; surge pricing can vary 5–6 times regular fares | Competitive; surge pricing only two times the regular fare |
| 6. | Innovation | ATG has a lot of technology muscle—its own proprietary AI software | Less room for investing—relies on partners; app proprietary but AV software from partners |
| 7. | Key differences | Uber also in meal delivery from restaurants and truck automation | So far only in ride-hailing |

*Table 8.2: Uber/Lyft comparison*

## 8.4 Ride-Hailing and TaaS Providers: Didi Chuxing

Didi Chuxing is the largest ride-sharing company in China, having amalgamated several smaller companies and acquired all of Uber's Chinese business in 2016. It provides transportation services, including taxi-hailing, private car–hailing, Hitch (   ial ride-sharing), DiDi Chauffeur, DiDi Bus, DiDi Test Drive, DiDi Car Rental, DiDi Enterprise Solutions, DiDi Minibus, DiDi Luxe and bike-sharing via a smartphone app, to more than 450 million users across over 400 cities in China.

*Figure 8.1: Chinese ride-hailing service Didi Chuxing (Source: Bloomberg)*

According to TechCrunch and CNN Money, Didi raised approximately $5.5 billion in 2017 at an valuation of $50 billion. In 2016, it had revenue of $7.3 billion. Its investors include Apple, SoftBank, China Mercantile Bank, Baidu, China Life and Alibaba.

Didi has set up a California-based research lab dedicated to developing artificial intelligence and self-driving car technology. It hired a senior Uber researcher to lead this research effort in the United States. It has also expanded into Latin America, via a $100 million investment.

Didi seems to be as ambitious as Uber in its intentions to develop AV software and full-fledged AVs that it can use for its fleet in the future in North America, where replacement of human drivers with robo-chauffeurs is the objective of several ride-sharing companies.

## 8.5 Ride-Hailing and TaaS Providers: Ola

Ola is an Indian online ride-hailing network. It started as an online cab aggregator in Mumbai and is now based in Bangalore.

Ola Cabs was founded in 2010 by two young graduates of IIT Mumbai, Bhavish Aggarwal and Ankit Bhati. As of 2017, the company had expanded to a network of more than 600,000 vehicles across 110 cities.

In March 2015, Ola Cabs acquired a Bangalore-based taxi service TaxiForSure for about US$200 million. In the same year, Ola acquired Geotagg, a trip-planning applications company for an undisclosed sum. Both of these acquisitions made Ola a strong force in the Indian ride-hailing market. Earlier, Ola had started an on-demand auto rickshaw service which is quite popular in India. It offers that service in 73 cities in India.

Like all other ride-hailing companies, Ola offers a whole range of services – from bare-bone economical service to a luxury service. As cash is still a common method of payment in India, Ola accepts both cash and its own credit card (called Ola money) payment. Ola provides different types of services, ranging from economic to luxury travel. Ola has 50-60% of the Indian ride-hailing market. It claims to service one million bookings per day.

According to Business Today India," Ola has raised USD1.1 billion in a new round of funding in October 2017 led by China's Tencent Holdings and existing investor Japan's SoftBank Group, and claimed to be in advanced talks to raise another $1 billion from other investors. The money comes as a booster shot for the homegrown start-up to keep up its intense competition with American rival Uber as both firms are burning millions of dollars every day in incentives and discounts for India's $12 billion taxi market."

## 8.6 Car-Sharing Companies: Car2Go

Car2Go is a German company (a subsidiary of Daimler AG) that offers a simpler, more convenient form of car rental than traditional rental agencies, in European and some

North American cities. Car2Go uses a smartphone app that really provides an early form of TaaS. It allows a customer to access (rent) a car anywhere, use it anywhere for as long as they need to and then leave it anywhere on the street in a designated zone, paying only for the time and mileage used. As the company's marketing materials say, "No reservations required. No long lines. No refueling. No worries." Cars are available on the street and in designated lots around each city, and users consult a live map on the app to locate the nearest one. Daimler wants to use Car2Go for a version of TaaS offering in Germany.

*Figure 8.2: Daimler's electric autonomous smart car from Car2Go*

## 8.7 Car-Sharing Companies: ZipCar

Zipcar is a car-sharing service like Car2Go that offers services through a smartphone app. ZipCar also has ambitions to start using autonomous cars in its fleet in the future. ZipCar could compete with Uber and Lyft or it could even be subsumed by Uber or Lyft or Didi. ZipCar has one advantage—it knows where the demand is concentrated; that knowledge would help it locate its vehicles close to the areas where demand for TaaS would be strong.

## 8.8 Fleet Operator Companies: Navya

Navya is a French start-up, established in 2014, that is focusing on autonomous fleets for geo-fenced and specified routes. Navya had a team of 120 employees in Lyon and Paris. It has also set up an assembly plant in Michigan for the North American market. Also, Navya partnered with Michigan University (called MCity) to test its fleet application in the university campus.

*Figure 8.3: Navya's campus fleet service (Source: Navya and the University of Michigan)*

Most analysts agree that it's easier to develop a fleet application of AVs because the routes for fleet operations are restricted, are not as congested and can be controlled far more easily. Lane markings, signs and traffic lights can be implemented by the operators so that there are fewer chances of accidents. In fact, I think it'll be fleet AVs that will lead the way toward fully autonomous level-4 and -5 vehicles.

According to Navya website[30], "In October 2015, Navya launched the Navya Arma, a 100% autonomous, driverless and electric shuttle that can transport up to 15 people and which is now operating on various sites using both private and public roads. The vehicle is equipped with numerous sensors and an on-board computing system that enables it to interact with its surroundings. It is also intended to provide efficient mobility solutions in terms of space and energy.

ARMA shuttle can reach speeds of 28 miles/hr. The vehicle aims to provide a complement to current public transport systems over distances that are too short to go by car or are too long on foot"

### Applications for Navya-like Fleet AVs

There are a number of excellent possibilities for where Fleet AVs could be successful, including:

- Industrial sites: transporting employees from public transport stations and parking lots to the work area
- Airports: transporting passengers and crews with their luggage to hotels or other airports
- Amusement parks: transporting visitors between attractions, parking lots and catering areas
- University campuses: transporting students
- Convention centers: transporting visitors
- Sports stadiums: transporting sports fans

## 8.9 Ride-Sharing Shuttle Service Companies: Tesloop

Tesloop is an interesting ride-sharing start-up by an 18-year-old Californian entrepreneur named Haydn Sonnad who wanted his father, Rahul, to fund his Tesla purchase. Instead, Rahul suggested that he come up with a business idea to fund his Tesla. So Haydn did—he proposed a ride-sharing service that would shuttle people between Los Angeles and Las Vegas in a brand-new Tesla Model S for $59 per person. The father-son duo did their research on costs, built some Excel scenarios and concluded that Haydn could make $5,000 per month if their projections were valid. So they launched a "Ride in red hot Tesla from Los Angeles to Las Vegas @59 per person" service. Tesloop offers a pristine luxury environment with in-car Wi-Fi, organic snacks and healthy drinks – driver provided or one of the passengers becomes a driver. With a novel business idea, free business press, and all kinds of endorsements from tire manufacturers and others, the service took off. The business started in 2016 and has now started services to other destinations. While the long-term viability of Tesloop might require further validation, Haydn's unique idea, endorsements from clean-tech enthusiasts and entrepreneurial drive have made him a ride-sharing star.

Tesloop is not only interesting in how it was launched but it's also a validation of certain TaaS trends with AVs:

- As RethinkX pointed out in its TaaS report, Tesloop has been driving Tesla electric cars for 17,000 miles a month—200,000 miles annually. TaaS enthusiasts think that the car may last for five years at this rate of use. Tesla offers a one-million-mile warranty even though, I believe, that it hasn't consciously designed the Model S to withstand such a heavy shuttle use. Tesla could introduce additional reliability features at additional cost after it has accumulated better data for battery life and replacement frequency.
- Tesloop has been using "free" Tesla charging stations. But Tesla may charge for this service in the future. It's expected that the cost of charging will work out to around 4–5 cents a mile.
- Tesloop's maintenance costs appear quite attractive. While a single car's maintenance record is not statistically significant, Tesloop spent only $11,000 in its first year use ($6,900 on scheduled maintenance and $3,500 for lamp replacement). This translates into 4–5 cents per mile.
- RethinkX's report says Tesloop's operational costs per mile (depreciation, finance, maintenance, fuel and insurance, but excluding driver costs) are in the 25-cents-per-mile range on limited short-term data. Adding in driver costs (two shifts at $30,000 per year per shift), it would become 55 cents a mile at minimum. At this cost, luxury autonomous Tesla cars become attractive from a TaaS point of view.

## 8.10 Fleet Operators: RideCell

RideCell is a San Francisco–based start-up that provides software to power TaaS, including car-sharing, ride-sharing, and fixed-route and dynamic transit services. Key investors include BMW i Ventures, Khosla Ventures and Y Combinator. It had a staff of 100 people and it acquired a small start-up called Auro, which was created by IIT and CMU alumni.

According to a company press release, "RideCell's unique differentiator is 'autonomous fleet operations'—technology that automates end-to-end business operations from consumer mobile apps to the day-to-day fleet management, demand and supply analytics, marketing, CRM and payments. Its mobility software platform automates some of the operational tasks in running a transportation system enabling car-rental companies, small municipalities and university campuses to launch on-demand, car-sharing and fixed-route services in weeks."

In addition to auto OEMs like BMW, RideCell's clients include transit agencies such as the Santa Clara Valley Transportation Authority in Silicon Valley, companies such as 3M, universities such as UC Berkeley, and hospitals such as UCSF Medical Center.

**Summary**
*I've explained the TaaS concept here, listed its cost elements and compared the economics of TaaS versus private ownership in a table—based on my assumptions. Business models are not static—the industry can certainly control costs of private ownership and manipulate business models. The industry may introduce TaaS at a loss (or low margins), getting us comfortable with the service, and then start increasing the price of TaaS when we are accustomed to the convenience of TaaS. As for our grandchildren, they may not ever learn to drive and may prefer the convenience of TaaS.*

---

**Citations for External References**

[30] Navya AV Fleet – https://navya.tech/en/navly-is-celebratng-its-1st-year-in-lyon-confluence-france/

# Chapter 9

# AV Perspectives, Part 1 (Consumers, Suppliers, Academics, Investors and More)

*In a business world, there are multiple players who participate in providing the end product or service to the consumer. Each of these players has its own perspective and objective, leading to individual perspectives and strategies. If everybody were pulling the rope in the same direction, we would move forward faster. But that's not what happens. Some differences are healthy for a better outcome. These differences in individual perspectives do slow the business down.*

We saw in Chapter 7 that there are a number of stakeholder camps in the AV ecosystem. Each category of stakeholders has its own perspective. On the surface, there may appear to be a common theme, but underneath that, there are divergent motivations. Here, I'll explore these perspectives so that we can try to appreciate what each group is trying to achieve.

Before we discuss those different viewpoints, let us look at where there is some agreement and unanimity. Almost everybody—consumers, suppliers and regulators—agree that there are too many accidents, injuries and deaths as a result of human-driver error. Auto manufacturers have been introducing safety features for some time. Seat belts, air bags and similar features have helped a lot. Over the past few years, luxury cars have begun adopting ADAS features, which help make driving safer. Safer, though not autonomous. Some regulators would like to prescribe some ADAS features as standard in all cars—from the inexpensive category to the luxury category where these features are sold as add-ons for higher margins. I see ADAS features trickling down into semi-luxury category first and then into less expensive cars over the next five years. Safety features and "look and feel" features are two ways of up-selling on a car. The ultimate feature that OEMs want to offer, of course, is the "autonomous" feature.

We can ask the following questions:

- Do consumers really want self-driving cars? If yes, do they want them for fun or they want to be freed from the necessity of driving?
- *Who*—among the passengers and drivers from different generations—really wants an autonomous car?
- What is the real motivation behind the AV innovation? Is it really innovators' altruistic desire to reduce deaths on roads, or is the potential reduction in road casualties just an argument to get these vehicles approved by regulators, so that

- OEMs can offer TaaS now and private ownership later for bigger and bigger profits?
- Are the established OEMs pushing AVs as the industry's answers to safety, environmental and congestion issues or are they just jumping in to prevent the Silicon Valley–based newcomers from taking over the auto industry?

I'll try to reflect on, and hopefully answer, these questions in this chapter from the perspectives of different stakeholders of the auto ecosystem. Let's start with consumers, for whom autonomous vehicles are meant in the first place.

## 9.1 The Consumer Perspective

There are a number of studies by independent, respectable organizations suggesting that the majority of consumers are interested in cars being made safer through ADAS features and maybe a moderate level of autonomous capabilities (perhaps at SAE level 2 or 3), but not SAE level 4 and level 5 where control is left completely in the hands of autopilot and where pedals and brakes have been taken away. Let's look at a couple of these studies.

The MIT AgeLab and New England Motor Press Association (NEMPA) conducted a survey, first in 2016 and then updated it in 2017, exploring consumers' perceptions and willingness to accept varying levels of automation. The 2016 survey of over 3,000 people found that while approximately a third of younger adults (under 45 years old) were somewhat open to full automation, older drivers were more likely to only endorse systems that assist the driver and that don't require them to give up control completely.

The study's findings show that despite significant progress in automation technologies (hardware and software) and media coverage, consumer confidence in fully autonomous vehicles dropped in 2017 as compared to 2016. Figure 9.1 shows the overall results across all participants, and Figure 9.2 shows the breakdown of responses by age group, with both showing the differences in results from 2016 to 2017. While the 2017 study still found that younger generations were somewhat more willing to accept fully autonomous cars than the older generations, almost half (48%) of all respondents said they would not buy fully autonomous cars where they had no control over the driving. This can likely be attributed to Tesla's crash in Florida and other news coming to light, showing that autonomous cars *do* get into accidents and that they may not be as safe as AV evangelists portray them to be.

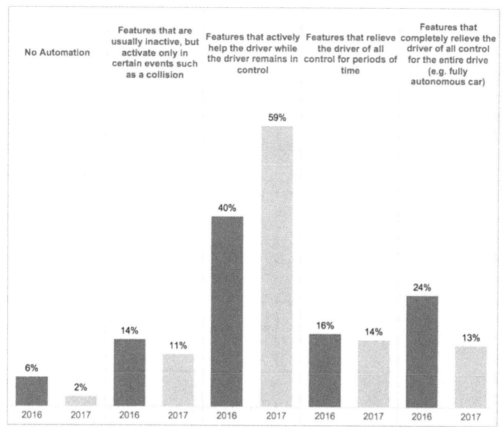
Figure 9.1: MIT study into consumer preferences of AV levels, 2017 (Source: MIT AgeLab)

## Maximum level of automation drivers would be comfortable with?

| 2016 | | 16-24 | 25-34 | 35-44 | 45-54 | 55-64 | 65-74 | 75+ |
|---|---|---|---|---|---|---|---|---|
| | None | 12% | 8% | 10% | 6% | 5% | 4% | 3% |
| | Emergency Only | 18% | 11% | 16% | 16% | 15% | 12% | 17% |
| | Actively Help | 27% | 25% | 21% | 41% | 44% | 56% | 52% |
| | Partial Control | 16% | 15% | 19% | 13% | 17% | 14% | 15% |
| | Full automation | 26% | 40% | 34% | 23% | 19% | 14% | 13% |

| | More comfortable with full automation | | | Comfortable with active assist, but not with giving up control | | | |

| 2017 | | 16-24 | 25-34 | 35-44 | 45-54 | 55-64 | 65-74 | 75+ |
|---|---|---|---|---|---|---|---|---|
| | None | 0% | 3% | 4% | 3% | 2% | 2% | 1% |
| | Emergency Only | 24% | 15% | 11% | 13% | 10% | 10% | 10% |
| | Actively Help | 46% | 43% | 49% | 55% | 63% | 64% | 69% |
| | Partial Control | 16% | 19% | 15% | 14% | 13% | 14% | 10% |
| | Full automation | 14% | 20% | 21% | 15% | 12% | 10% | 10% |

As compared to 2016, all age groups are less interested in full automation. Effects are particularly pronounced among younger respondents.

*Figure 9.2: Maximum level of automation drivers say they would be comfortable with (Source: MIT AgeLab)*

In an interview with Zeninjor Envemeka of Radio Boston in 2017, Bryan Reimer, a research scientist at MIT's AgeLab, suggested that consumers' responses are based on their experience with technology in general—struggles with Wi-Fi, the internet going down and dealing with automated help centers. And now "you're asking an individual to put their life in the hands of technology," he said. "They're looking for driver-assistance systems that work to help them stay in active control [and] safe control of the vehicle. They're just not looking for a car to drive them in a chauffeur kind of framework." He suggested that those working on autonomous vehicles should build consumer trust through public disclosures and investigations into vehicle failures.

Studies by J.D. Power and the University of Michigan support the MIT study's conclusions that the majority of consumers are interested in ADAS features but are not so sure about giving complete control to the robo-chauffeur.

A survey of 505 drivers by the University of Michigan's Transportation Research Institute found that most participants weren't particularly happy about the idea of automated cars, "with the largest segment of respondents indicating they'd prefer to retain full control over their vehicles."

The most frequent preference given for vehicle automation was for no self-driving capability at all (43.8%), with partially self-driving vehicles as the second preference

(40.6%). Just 15.6% of respondents said they liked the idea of a self-driving car being fully in control.

Andrew Moore, computer science dean at Carnegie Mellon, told The Atlantic, "No one is going to want to realize autonomous driving into the world until there's proof that it's much safer, like a factor of 100 safer, than having a human drive."

In 2016, Deloitte, an enterprise consulting company, conducted an extensive survey among 22,000 consumers in 17 countries to find their preferences for increased automation of the driver function in cars, including the trend toward fully autonomous cars. Deloitte's findings are generally in sync with the other studies, as you can see in Figure 9.4.

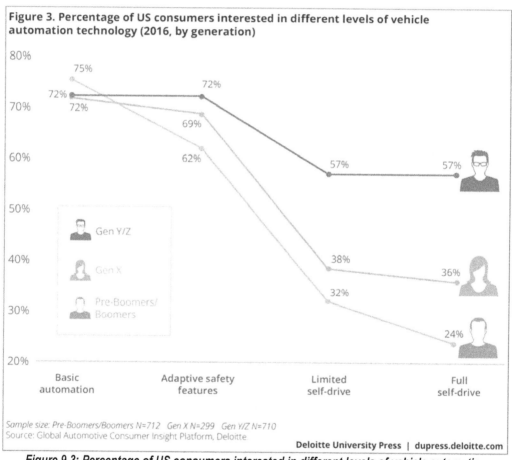

*Figure 9.3: Percentage of US consumers interested in different levels of vehicle automation technology (2016, by generation) (Source: Deloitte study)*

To summarize the key findings from these studies:

1. Consumer interest, especially among the younger generations (Y and Z), in autonomous cars increased from 2014 to 2016, while the MIT survey showed

decreased interest from 2016 to 2017. The increase in the 2014–2016 period can be rationalized on the basis of extensive media coverage and relatively positive stories about technological progress made during this period. The decrease in 2017 can be attributed to the Tesla crash, accidents in AV trial cars and other news about human intervention in trials.

2. There is virtual unanimity among US consumers (in all generations) that safety-related features are of maximum interest. As the Deloitte study said, "Across all US consumer segments surveyed, features that improve driver and pedestrian safety are perceived as much more valuable than those that enable connectivity, comfort, or even fuel efficiency," including fully autonomy.

### 9.1.1 How Much Will Consumers Pay for Autonomous Features?

The Deloitte study also found that US consumers' willingness to pay for new automation features decreased over time—implying that suppliers (both OEMs and newcomers) will be under a lot of pressure to put forward a good story to support their case.

According to Deloitte, the additional value that consumers assign to automation features is quite low—in the $975 range per feature. This is significantly less than the cost of these features and what the OEMs are charging for them in the luxury-car category ($3000–$6000). The MIT survey asked respondents how much they would pay for an entire driverless car (as opposed to the delta increase for autonomous features), and found, as shown in Figure 9.4, that 48% of drivers said they would never buy an autonomous car at all!

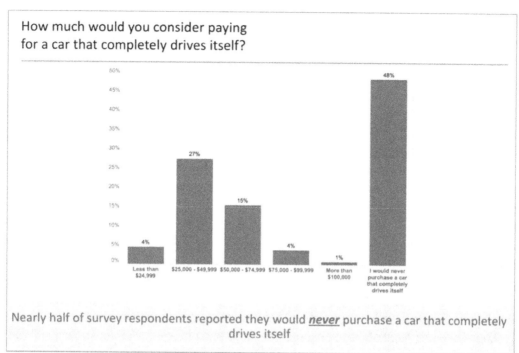

Figure 9.4: How much MIT-survey respondents said they would pay for a driverless car (Source: MIT AgeLab)

### 9.1.2 Consumer Perspectives toward TaaS

Many proponents of AVs suggest that individual vehicle ownership will decrease significantly. They believe that the future TaaS business model using AVs will result in a lower per-mile cost to most consumers, so it will be cheaper to use TaaS/MaaS providers than own a car.

While consumers seem to like the prospect of this, they still don't truly trust the cost projections. Financially educated consumers think that costs will go down only for a small number of consumers who drive a low annual mileage (<3,000 miles). Also there are a number of situations where consumers still believe ownership gives them more flexibility, convenience and timeliness than calling a TaaS provider like Uber or Lyft. I tend to agree with this assessment on the part of consumers, as I discussed in Chapter 8.

**Consumer Attitude towards Societal Benefits**
Major rational for AVs is that AVs are safer than human drivers. However consumers view personal safety at a higher priority than societal safety. That viewpoint emanates from fundamental human survival instinct whereby consumers would select options which benefit them over the option that benefits society at large.

**My Observation on Consumer Studies**
Remember that surveys measure respondents' general preferences at the time of the study, and in this case their anticipated preferences in an imagined or wishful future. If we asked "would you buy an AV if one were available tomorrow at this price?" we might get

different set of responses. Despite the results, I believe that these surveys suggest that consumers may be willing to adopt AVs in the distant future when they are affordable and customers have trust in their safety. For the near future, majority of them will be highly cautious.

## 9.2 Perspectives of AV Suppliers

There are two distinct categories of AV suppliers—the established auto OEMs and the newcomers from Silicon Valley. They may be chasing the same target and may be cooperating in some areas but their perspectives and strategies are quite different. Let's discuss their perspectives separately.

### 9.2.1 Auto OEMs

Auto OEMs were not the leaders or trendsetters behind the current race toward autonomous cars. OEMs are old, though strong and steady—they build reliable cars in the millions for their customers. They run efficient assembly lines and keep millions of people employed. They take well-calculated and risk-free steps before making changes to their product lines, and most of these changes are small and incremental. Every now and then, some trail-blazing product executive breaks ranks and introduces something significantly different but OEMs don't bet the farm on such an effort. They are micro innovators. They are from Detroit and Stuttgart—not from Silicon Valley.

GM's Futurama exhibit in 1939 and the Daimler-supported PROMETHEUS project in the 1980s certainly suggest there was existing *interest* among OEMs in exploring the AV world eventually, but would the Detroit-based crowd and their Stuttgart-Munich compatriots would have dared to announce on their own that there would be a self-driving car in 2021 or 2025, or implemented ADAS features in their luxury cars and invested in AV R&D so aggressively if it hadn't been thrust upon them by disruptors like Google, Tesla and Uber? It' is fair to say that auto OEMs have decided to get into the autonomous car race because disrupters from Silicon Valley are threatening their territory and they are fighting hard to protect their turf.

I reviewed future auto landscape reports from some of the world's largest and most well-respected consulting companies. They are preparing OEMs to adapt to the emerging environment. The following common perspectives emerged, whether that is what their advisers are telling them or that is what they have realized themselves:

1. **OEMs must change with the changing landscape**: The auto marketplace will change drastically over the next two decades. OEMs must upgrade their product lines, change their strategies, acquire internal expertise and enhance their competence in computer hardware/software integration. They must build relationships with multiple suppliers and change their business models to hybrid ones that are an appropriate blend of traditional auto hardware models and those employed by software/telecom/mobile app vendors of Silicon Valley.

2. **The AV-related market is too big and too lucrative to be ignored**: After an initial period of complacency (2010–2014), OEMs have woken up to realize that the AV-related market is too big to be ignored. AT Kearney, a management consulting company based in the United States, estimates that this market (autonomous cars plus the hardware required to enable them, vehicle-to-vehicle [V2V] mobile broadband communications, mobile apps, etc.) could be worth $282 billion by 2030—that's 7% of today's $3 trillion auto market. Of this, the market for autonomous vehicles themselves—the focus of this book and most observers—is expected to be $92 billion. After 2035, this market could increase to $560 billion—17% of today's total auto market. More importantly, the new AV-related market will have higher margins than the current car business, not to mention that margins are going down in the car-selling business. OEMs are being told by their advisory consultants that they must prepare for this new battleground consciously and deliberately. If OEMs decline to participate, not only they could lose out on this new business, it could erode their core business as well in the post-2025 period.

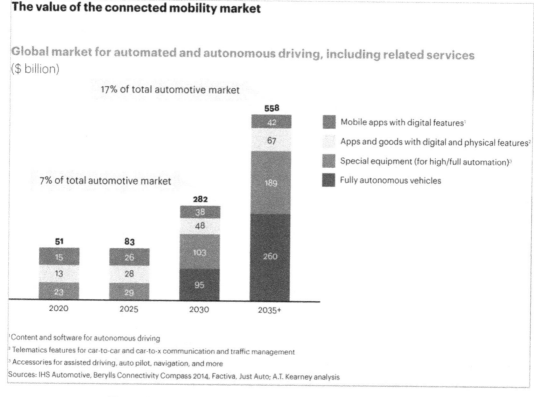

*Figure 9.5: The value of the AV market (Source: AT Kearney)*

The growth from 2030 onward will not be uniform across all categories—the market for specialized hardware to make AVs (ADAS and its future enhancements) will peak around 2035 (after which it will become a commodity), and revenue from mobility content, software and services will exceed that from specialized ADAS equipment.

3. **The road to fully autonomous vehicles will be gradual:** Most independent analysts and OEM executives agree that autonomous vehicles will not arrive on a specific date in the future. Instead, OEMs believe that AVs will arrive in a phased fashion in several gradual steps. It's also not clear whether regulations, insurance and liability issues will force OEMs into such gradual steps, such as offering AVs on a TaaS basis where they take on the risk themselves in order to ease us into acceptance of the robo-taxi concept. If that happens, they may delay offering AVs to the general public for purchase, even though there are individuals ready to order them. Much of the OEM perspective is based on this gradual arrival of AVs.

4. **Established OEMs are not willing to concede to Silicon Valley challengers:** In spite of the technological and entrepreneurial advantage that the challengers from Silicon Valley may have over the established OEMs, OEMs are not ready to cede the market they have built over the past hundred years. They have their own unique strengths—seasoned management, automated and finely tuned assembly lines, a trained workforce, and dealer and service networks. OEMs are doing what is necessary to protect their turf—establishing R&D centers in the heart of Silicon Valley (Palo Alto in particular), investing in or buying up auto-centric start-ups and established companies in AI, mapping software and ADAS hardware, striking close partnerships with the leading AV technology companies and making AV strategy a high-priority component of their future strategy. Most OEMs have announced a timetable for introducing autonomous cars around 2021, with a disclaimer that puts the responsibility on regulators to clear the way. They have also accelerated their implementation of ADAS features in both luxury and semi-luxury cars. More affordable cars will also not be left behind, they say (the Ford Fusion and the Chevy Bolt, which have both been used in AV trials, are not Ford's and GM's most expensive brands). The OEMs are courting both Uber and Lyft for TaaS, while preparing to enter the TaaS market directly on their own. They certainly want a piece of the TaaS market, especially if the widespread adoption of AVs sees a real decline in individual ownership as projected by market forecasters, especially among Y and Z generations. In a sense, OEMs are betting their chips on all the horses that are expected to hit the tracks.

5. **The current stock market infatuation with Silicon Valley auto kids is not fair to the auto OEMs:** The stock market has changed significantly over the past 40 years with the stocks of mature, dividend-giving companies like the auto OEMs being valued more on their current performance and those of the new, entrepreneurial Silicon Valley AV companies being valued more on their future potential and media backing. Therefore, auto executives feel that Tesla's market valuation, which several times in 2017 exceeded the valuation of GM and Ford, is irrational exuberance and

inconsistent with the company's ability to meet production targets and become profitable. Unfortunately for OEMs, the stock market and adulation from the technology-influencing consumer crowd have the upper hand. It's in this context that one former GM board member made a statement in November 2017 that Tesla was going bankrupt. The stock market paid no attention to his pronouncement. Fair or not, the autos OEMs do realize that the stock market and investors are backing the Silicon Valley newcomers.

6. **OEMs believe in both ICE and electric AVs**: ICE vehicles are OEMs' bread and butter. Their manufacturing lines and supply chains have been fine-tuned for ICE vehicles. After a lot of pressure from regulators, they're struggling to develop electric versions, reluctantly but surely moving toward electric vehicles at a pace they feel comfortable with.

7. **Some OEMs are still reluctant to adopt Apple CarPlay and Android Auto:** While some OEMs have started offering Apple CarPlay and Android Auto integration in some of their models, they've done so unwillingly, and it doesn't mean they're giving over the smartening of the AV to Apple or Google. They want to adopt these tools such that their own infotainment portion of the OS controls the integration. I shall comment on this further in later chapters.

8. **OEMs feel that they have the wherewithal to deliver AVs in quantities the market needs:** OEMs feel that they have the assembly-line infrastructure, the experience and the workforce to deliver the bodies of AVs in the massive quantities that the market needs. The challengers cannot scale up quickly. Look at Tesla's experience in terms of keeping its delivery promises for Model 3.

## 9.3 Perspective of Silicon Valley Challengers

In this category, we have Alphabet Waymo, Tesla, Uber, Lyft and the AI-focused vendors. These folks are visionaries, innovators and entrepreneurs. They only look at the future. They are believers in change and they try to change the world, creating new markets out of nothing. They think that it's time that our transportation methods change, and they have the solutions for that change. They will write the rules of this change, putting OEMs at a competitive disadvantage unless OEMs follow the challengers' rules. Their perspective is different from that of the OEMs. They give credit to OEMs for running efficient assembly lines, and do not want to give up their dominance of high-margin software components and applications. The members of this group tend to share the following beliefs:

1. **Level-5 autonomy is the right way to go:** The leader of this group Waymo group believes in the target destination: fully autonomous cars. Waymo believes that semi-

AVs are not good enough. They want the real deal—SAE level 5, ASAP. Tesla is promising fully autonomous vehicles but as the end target. Silicon Valley believes that problems are created when human drivers, who are preoccupied with other tasks, try to take over from autopilot. There could be a significant delay when the autopilot in a level-4 car asks the human driver to take over and the human has to figure out what to do. The human driver could take too much time or even take a dangerous step. And in a TaaS scenario, there would always be the question of which passenger was the designated "take over" driver. For these reasons, this group believes level 5 is ideal.

2. **TaaS is coming fast, and will come before individual ownership**: Should AV manufacturers target private ownership of AVs or just offer TaaS for a fee? The Silicon Valley group believes that apart from technologically inclined eager beavers, there may not be a huge crowd waiting to buy AVs without trying them out for more than an hour-long test drive. They'll want to experience it as a service first through TaaS providers. They also may not want to make the financial commitment for AVs. Younger generations (Y and Z) also don't have the ownership bug the way baby boomers do, and they certainly don't see owning a car as a status symbol. To them, transportation is a matter of necessity more than anything. As we saw in Chapter 8, TaaS advocates believe that the new service is expected to be cheaper than owning a car. Therefore, the Silicon Valley group (and OEMs to some extent, too) believe in the TaaS business model.

3. **Driving habits should change disruptively—not slowly and incrementally**: The Silicon Valley challengers believe that as soon as fully autonomous vehicles are available, drivers should become passengers—in fact, there should no longer be any human drivers. They believe that human passengers should be allowed to give direction to the AVs but not to control the actual driving.

4. **Consumers will benefit from integrated connected mobility**: This group will offer integrated connected mobility, which will involve a supercomputer–based operating environment that supports the following functions:
   - A single operating framework that controls everything in the AV
   - An autopilot handling the core driving task
   - Integration with the car's infotainment subsystem
   - Support for solutions like Apple CarPlay and Android Auto
   - Connectivity with other cars in the vicinity (V2V) and with the road infrastructure (V2X)
   - Support for personalization for passengers or AV owners
   - OTA maintenance support and a help function for troubleshooting

5. **They have an edge over OEMs in building AI solutions for AVs:** The core competency of this group is in building businesses around computer software and hardware. They have the IT-based know-how that's required to convert 3,000 pounds of steel/aluminum/plastic machinery into a robot that will drive people around.

## 9.4 The AV-Following Media

There has been extensive coverage of self-driving cars in the media—mostly after Google started its self-driving car project in 2009. The idea seemed like something out of a science-fiction story—many people may not have believed it was true when they heard it for the first time. Gradually the news sank in as the public started seeing AV trials on public roads. The concept simply excited the imagination of journalists and they wrote enthusiastically every day about one aspect or another of the AV industry. The coverage was rarely holistic, each story representing only one viewpoint of the bigger story. One had to piece together many stories to get a comprehensive understanding of what this might mean for us as a society.

Technology journalists provide a great service. However, in many cases, they are not subject matter experts—they deliver the news to us but do not provide expert and validated opinions of the future. Perhaps that is not their job. The press is hungry for "newsworthy" content—anything that makes headlines, is significantly exciting to readers and that goes against conventional thinking. News about incremental changes does not excite people—that's business as usual—but disruptive technology makes great press material. Journalists are not serious analysts or consultants who look at all the factors and angles before rendering a piece of advice or guidance. They tend to extrapolate and sometimes overshoot what's possible in the immediate future. Most of the information journalists write about is provided to them by OEMs' and technology companies' PR staff. The PR team's job is to put out the most optimistic picture without mentioning the assumptions or explaining the challenges their companies face in delivering. Most of the timeline forecasts by vendors are too optimistic.

Over the past several years, I've reviewed over 1,000 AV-related articles and reports from various organizations (freelancers, bloggers, big-5 consulting companies, reputed media such as *MIT Press, Wired* magazine, Bloomberg, Reuters, Business Insider,fastcompany and TV). My general perception is that coverage on autonomous vehicles is extensive—exceeding the coverage for any other technology innovation from the past 20 years except perhaps the smartphone. The coverage provides a good representation of what's going on in the industry. The majority of the articles on the web are about announcements by the OEM/technology vendor community or interviews with marketing executives from these organizations. Opinions vary a great deal across the whole spectrum. One thing's for sure—based on the opinions of journalists, you can't come to any easy conclusions. To develop my own educated opinions, I relied more on independent sources that were not aligned with any vendor and my professional insight. I looked for cross-validation.

The media's perspective on AVs is *let's create the hype, carry the message to the reader/viewer, get heard or read and let the audience form its own opinion; we are messengers and our job is to carry the message.*

## 9.5 Perspectives of Academia and University Researchers

Academia and university researchers are among the most important players of any technology ecosystem. Many universities in Europe and the United States have been involved with AV research for the past 40 to 50 years. Their efforts are bearing fruit now. Without their hard work, there would not be an AV industry today. In Chapter 3 we covered their work, which has had four major thrusts:

1. Targeting the safety issues of cars, leading to the design of ADAS features using electronic sensors, LiDAR and camera-based computer vision.
2. Developing electric cars (primarily by replacing the internal combustion engine with lithium batteries and replacing the mechanical transmission with direct drive).
3. Using artificial intelligence, especially deep learning, to make cars fully autonomous.
4. Considering softer issues related to ethics, urban transportation planning and the impact of AVs on cities.
5. Limited research into consumer attitudes.

Here's my take on the perspective of the academic and research community on the AV conversation:

1. Using AI in AVs is incredibly complicated. A lot of work still has to be done before we reach our goal of CMU's Sigma 4 or 5 quality control standards for this mission-critical application where matters of life and death are involved.
2. The academic community's consensus is that our goal of SAE level-5 AVs *will* be achieved in future. In level-4 AVs, drivers must stay alert or run the risk of occasional be catastrophic results, but MIT and Stanford University academics feel that human drivers may not take control in a reliable fashion in emergency situations.
3. The community does not endorse the industry's timeline for introducing AVs to the public.
4. It *is* enthusiastic about continuing advancements in AI, quantum computing and robotics.
5. The academic community is ambivalent about the ethics debate in the conversation.
6. It is supportive of a holistic approach to solving traffic congestion and safety issues.
7. Finally, the community agrees that its role is to provide raw code, often written in a hurry to prove that something can be done, but that the industry owns the

responsibility of optimizing it for widespread industrial use that may require CMU's CMM methodology and associated quality standards.

## 9.6 Perspective of Technology Innovators and the VC Community

In Chapter 2, we discussed the fact that academic researchers are usually the first group to invent new ideas, concepts and technologies. However, it's the technology innovators who give life to these ideas because they are constantly investigating untapped discoveries to create products that are useful to society. The innovators in the auto industry are more enthusiastic about autonomous cars than anyone else. These innovators came out of California's Silicon Valley, Germany's auto R&D labs and Israel's scientific community. Their perspective is *full steam ahead with AVs*—all the way to SAE level 5.

A huge amount of venture funding and investment has been pouring in from both traditional capital markets and OEMs themselves. While it's difficult to accurately state the amount of investment, Brookings Institution estimates that between 2014 and 2017, AV investments exceeded $80 billion. This trend will continue to accelerate in 2018 and beyond. Figure 9.6. shows Brookings' listing of major investments in the industry; the more well-known among them include:

- In 2015, Audi, BMW, Daimler and other German automakers acquired digital mapping company HERE for $3 billion.
- In August 2015, Hyundai and Kia invested $2 billion in artificial intelligence.
- In Jan 2016, GM invested $500 million in Lyft.
- In March 2016, GM acquired Cruise Automation for $1 billion.
- In May 2016, Apple invested $1 billion in Didi.
- In August 2016, Uber acquired Otto for $880 million.
- In January 2017, Intel invested $390 million in HERE.
- In February 2017, Ford invested $1 billion in Argo AI.
- March 2017, Intel acquired Mobileye for $15.3 billion.
- Hundreds of millions of dollars of VC money has been injected into start-ups all over Silicon Valley.

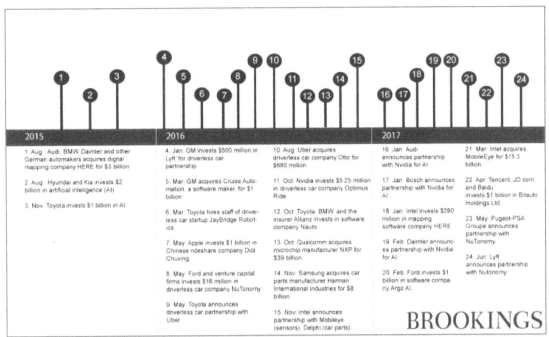

Figure 9.6: Prominent AV investments (Source: Brookings Institution)

The perspective of the innovators' and VCs is that AVs are for real and the transportation industry is going to change like never before. Trial results prove that this one is not a fad. With so much money at stake, so many start-ups, so much brain power from Silicon Valley and so much manufacturing experience from Detroit and Stuttgart, we are sure to see a huge transformation (says this group).

## 9.7 Perspective of AV Evangelists and Early Adopters

AVs are exciting and advance the state of the art in computer control of cars. They're creating a new paradigm. AV-related AI will truly break new ground and open it up for many other applications in the future. Therefore, there's a growing group of early adopters and trendsetters who are waiting for fully autonomous vehicles to show up in dealer show rooms. There's one primary aspect of AVs that matters to them: the ability to perform their favorite tasks while they are being transported in their AV. Nobody is sure of the size of this group, but there's no doubt it exists. These folks' perspective is that AVs will solve many of the traffic congestion and safety problems that cars with human drivers have created in the cities of the world. They want AVs to be approved by regulators sooner rather than later.

## 9.8 Perspective of Infrastructure Owners—Cities, States/Provinces, Federal Agencies

AVs need roads to run on. Roads and highways are owned by cities, municipalities, states/provinces and federal agencies. Are these infrastructure owners ready for AVs? Some analysis has been done by universities (in Michigan, Texas, Ohio and California, as well as in Canada, Germany, the United Kingdom and Australia), independent transportation planning agencies (such as the Victoria Transport Policy Institute of British Columbia, Canada) and the American NHTSA itself.

I will deal with regulatory issues separately in Chapter 10, but I will briefly mention the opinions, concerns and perspectives that these bodies have with respect to AVs and transportation planning.

1. Cities are enthusiastic toward AV trials: Trials give cities good press and they like their cities to be portrayed as progressive and welcoming to new technologies for economic development purposes. In this context, Las Vegas, San Francisco, Orlando, Phoenix, Paris, London, Stratford (Ontario), Edmonton (Alberta) and many others have welcomed the setting-up of test tracks or have zoned specific areas for AV trials.
2. Cities expect there to be a mix of AVs & non-AVs for several decades: There are over a billion vehicles (including both passenger and commercial) in the world, with 230 million in the United States alone. AVs will initially start dribbling onto the roads in small numbers, as the rate of adoption will depend on a number of factors (as discussed throughout this book). Infrastructure planners expect that this process will be slow and gradual because of the great number of unknowns. Some forecasts suggest 2050 as the tipping point when AVs will become widely adopted.
3. Industry forecasts are too optimistic: The author of a well-researched report by the Victoria Transport Policy Institute feels that many industry reports paint a rather optimistic picture of the benefits of AVs—parking spaces converted into parks, less congestion in the cities, less pollution and reduction in death rates. Another study, by Le Vine, Zolfaghari and Polak at Imperial College London in 2015, even suggested that AV passengers may want slower acceleration and speed than today's impatient human drivers, which could result in *slower* traffic and *reduced* capacity on roads and highways.
4. Cities are not ready for AVs: City planners say that while the industry is ready to build AVs and is working hard at it, the cities are not ready. For AVs to operate safely, cities need to upgrade their infrastructure, including roads, lane markings, road signs, signal lights and so on. Cities will need huge capital expenditure to do this. Can they continue to fill pot holes and upgrade at the same time, without increasing taxes. Public is not willing to accept more taxes. There's no easy answer
5. Cities would need more tax dollars if AVs replaced today's cars: If the number of registered vehicles decreases and the taxi industry goes out of business, cities would lose an important revenue source. The cars would use fewer parking spots, which are also often owned by cities. Increasing the number of AVs would require major upgrades to the road infrastructure. All of this would require cities to increase their property taxes or apply a special AV tax—certainly for TaaS providers.

6. We need uniform coordinated national regulations for AVs: Cities are not ready with policies, regulations and by-laws to manage the onslaught of AVs. Individual cities may choose to implement their own rules but that may not be in the consumer's best interest. Cities and consumers want a uniform set of regulations and by-laws that would govern the use of AVs across jurisdictions.
7. Cities advocate for holistic solutions to combat congestion and auto-related casualties: City transportation planning experts believe that there are multiple ways of reducing car accidents and congestion. AVs might be one of those tools but the cost may not be in the best interests of society at large. Independent transportation research experts need to investigate this problem in a holistic manner.

**Summary**
*I have presented several distinctly different perspectives of major groups within the AV ecosystem. Consumers are in no big hurry to ride in AVs. Technology challengers want to get there quickly in one big step. OEMs want to reach the destination in several incremental steps, stopping at check points and recalibrating their strategy, if necessary. The academic community is cautioning that we need more research, particularly into AI. And cities are in no great hurry as they are not ready for AVs.*

# Chapter 10

# AV Perspectives—Part 2 (Insurance, Legal, Regulatory and Ethics)

*There are perspectives galore in the AV conversation. I covered suppliers', consumers' and promoters' perspectives in the previous chapter, among others. However, there are more perspectives to consider—insurance, legal, regulatory and ethics. Let's focus on these here to complete the picture. The insurance industry is confused and concerned. Lawyers are not sure that existing laws serve the AV industry adequately. Regulators are paying attention but taking a passive role as if the problem will be resolved by the industry. Finally, the ethical questions surrounding the AI autopilot are creating significant unease among the civil society.*

## 10.1 Insurance Issues

I've said it many times throughout this book: the most important argument in favor of taking the control away from accident-prone human drivers and giving it to AI-based autopilots is that AVs are safer and will decrease the number of car accidents, reduce personal injury and save lives. The insurance industry agrees that the majority of accidents are caused by human error—driving too fast, drunk driving and, more recently, distracted driving, among other things. The corollary of these expectations is that if AVs lead to fewer accidents, the cost of insurance should go down. Indeed it should—that's what industry analysts, including those from the insurance industry, agree on. In a recent interview with CNN, Warren Buffett, head of Berkshire Hathaway, told CNBC's *Squawk Box* that "if autonomous vehicles prove to be safer than regular cars, insurance costs will plummet." Consumers wonder whether premiums will really come down or just get hidden somewhere else. The second question is *when* premiums will come down, because actuaries in insurance companies are a very conservative group; they only reduce premiums after collecting a lot of historical data. That takes time.

There are two reasons for insurance costs to come down:

1. First, the actual costs of property loss, personal injury and liability claims will come down because of the reduced number of accidents.
2. Second, auto companies themselves will take on some risk, offering a bundled price that shifts car insurance costs from the owner to the manufacturer. In fact, Tesla has started doing exactly that in Asia where it's including insurance coverage in the price of the car itself. The reason Tesla's doing this in Asia (and may do the same in North America in the future) is that after the installation of Tesla's Autosteer software, accidents involving its 2014–2016 Model S and 2016 Model X vehicles went down by

40% (see Figure 10.1). Whether Tesla decides to carry liability risks for its cars as a component of its corporate insurance plan with its insurance company or just treat it as a business risk, nothing reduces premiums like competition.

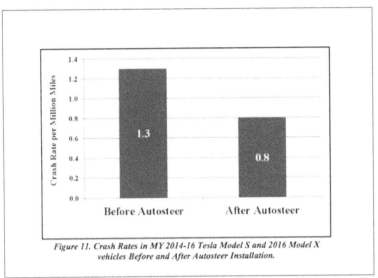

Figure 10.1: Reduction in accidents after installation of Tesla Autosteer (Source: NHTSA –Data Supplied by Tesla)

This phenomenon could hit the insurance industry hard. The personal auto insurance sector could easily shrink to 40% of its current size (which is $200 billion in the United States alone) within 25 years, according to a report by KPMG.

### Questions about Liability in Insurance Coverage

Legal experts are not sure whether existing laws can address the needs of AVs and are hoping for new legislation to clarify the ambiguity that exists right now. Some legal experts are taking a stab at the current liability legislation based on how AVs are expected to behave. Geoffrey Drake, a partner in King & Spalding's tort and litigation group, told Business Insider that liability responsibility depends on the type of car you're driving or are a passenger in. If you're driving an SAE level-4 or level-5 car without pedals or a steering wheel, the manufacturer will bear the responsibility because the passengers (including the owners of the AV) wouldn't play any role in driving the car. The argument in that case would be that the accident occurred because the car behaved the way it was programmed.

So, manufacturers may have to assume responsibility for self-driving software. If they use third-party software (e.g., if a company like Waymo supplies software to OEMs), the two parties may have to strike a legal agreement as to who will carry the insurance coverage. But whoever it is, it does appear that the passenger/owner of the AV won't have to worry about the liability.

A precedent was set by Volvo, which announced in 2015 that it will take full responsibility for the correct operation of the robo-software in its level-4 and level-5 AVs. This might have been a PR announcement without too much substance because Volvo has delayed its AV initiatives until 2021. Nonetheless, it remains to be seen whether other auto manufacturers will follow Volvo's lead. The AV vendors will have to figure out whether they will self-insure the risk or buy insurance in bulk from the reinsurance market.

The liability situation becomes a bit complicated with level-2 and level-3 AVs where there's a fallback function for the driver. Here, too, however, the precedent has been set—manufacturers clearly state that the driver must always be in control. ADAS hardware and supporting AI software is there to passively or actively assist the driver, but if the driver doesn't take over when the ADAS alerts him/her, the responsibility for any accident is nevertheless transferred to the driver or the owner as existing insurance laws permit. If, on the other hand, an accident is due to malfunctioning robo-software (say, it doesn't detect a hazard or slow down when the car in front of it slows down), this becomes a disputable situation that lawyers and insurance companies will have to figure out.

**Responsibility for Damages** *(Source - Insurance Institute of Canada)*
The Insurance Institute of Canada summarizes the situation neatly, as shown in Figure 10.2.

If you are driving a non-autonomous car, it is business as usual: you and your insurance company are responsible if you hold a valid paid-up policy. If you're driving a level 1, 2 or 3 semi-autonomous vehicle, you and your insurance are still responsible, although some responsibility may be transferred to the manufacturer if there's a malfunction of the AV-related autopilot software. If you're a passenger/owner of a level-4 or level-5 AV being driven by a manufacturer-provided autopilot system, it is the manufacturer's responsibility.

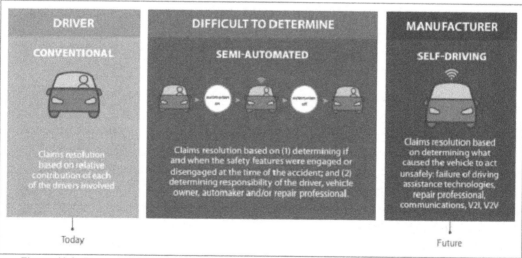

Figure 10.2: Responsibility for damages based on type of car (Source: Insurance Institute of Canada)

## The Tesla Crash in 2016

The Tesla crash in Florida in May 2016 provided a strong message to legislators that insurance and liability laws need clarification to make the public absolutely sure as to the driver/owner's responsibility in level-2 and level-3 AVs. The Tesla Model S (which could be classified as a level-2 AV) failed to brake because the car's autopilot hardware failed to see the white side of a tractor trailer making a left turn up ahead because of bright light from the sky. The photo in Figure 10.3 shows the terrible result.

Tesla issued a statement that its autopilot system worked as designed and that the driver should have been alert to take over—the autopilot system gave him seven seconds to apply the brakes. After an initial investigation, the NHTSA determined that the autopilot worked as designed and was not at fault.

*Figure 10.3: The fatal Tesla accident in Florida (Source: NHTSA)*

The Tesla crash demonstrated that the driver is still responsible for accidents and must hold appropriate insurance coverage.

### Factors Affecting Insurance Premiums for AVs *(Source - Thinkinsure.ca, 2016)*

The following factors may influence insurance premiums for AVs in future:

1. Insurance premiums will go down with AVs because of fewer accidents and fewer medical claims.
2. There will still be some accidents involving AVs, although fewer than under the current environment.
3. The cost of repairs for a single accident will be higher because of embedded sensors and the number of computers involved.
4. Cars will have a log of their driving—like the black box in airplanes—whose information may be used to resolve disputed cases.
5. Your prior driving record may not be a major factor in determining your premiums.

6. Manufacturers may assume liability for level-4 and level-5 AVs.
7. Insurance fraud should decrease over time, because there will be a lot of recorded data from both cars to determine the cause of the accident..

**What Needs to Happen**
1. Federal departments of transportation in different countries need to clarify their current legislation on car insurance and should ask legislators to amend existing laws.
2. Federal agencies (like NHTSA in the United States) need to persuade insurance associations/bodies to educate the public in terms of customers' and insurers' responsibilities. There should be no ambiguity.

## 10.2 Legal Issues

Technology generally moves ahead of the laws regulating an industry. In the case of AV, there are a number of legal areas where there is ambiguity based on the current set of laws governing drivers, vehicles and manufacturers. Use of artificial intelligence to replace a human being in operating a vehicle is a brand-new area that's raising concerns among lawyers and legislators. On the one hand, government administrators are encouraging the AV industry to conduct AV trials in their jurisdictions, mainly for economic reasons, but on the other hand, they aren't doing too much to draft new regulations and laws to manage the arrival of AVs. In the United States, the current federal administration (the Department of Transport -- DOT) is taking a "hands-off" approach to regulations, despite the fact that auto manufacturers want unified federal regulations to be drafted soon. Several US states (Nevada, California, Florida, Michigan, Tennessee and D.C.) have passed legislations that address some of the issues, not all. A number of US law firms have issued white papers and discussion documents to bring various legal issues to their clients' attention. In the United Kingdom, members of parliament have proposed a draft bill. Similarly, some European countries are in the early stages of discussing new regulations and/or laws. Here, I'll briefly outline some of the issues that regulators and legislators must resolve.

**Legal definition of a driver, a passenger and an autonomous car:** Autonomous cars may not have pedals or steering wheels and may only have an electronic console that accepts limited input from the occupants. If that's the case, the relevant DMV laws must clearly define who the driver is—is it the human being or the entity that owns the vehicle? What happens if the AV is individually owned by a human being who loans it to a third party (a family member, a friend or an employee)? There are a number of situations possible that are brand-new in the world of AVs. The definition of "autonomous car" itself is being defined by the SAE and the NHTSA. Relevant laws will have to specify which definition applies.

**Who's responsible for traffic violations?** Currently, most jurisdictions fine the driver of a non-AV for violating traffic rules. These laws will have to change because the responsibility for the violation will vary depending on the situation.

**"Safe harbor" provisions for after-market AV upgrades:** Some proposed laws for autonomous cars have a "safe harbor" provision that protects OEMs whose vehicles have been converted into autonomous cars by third parties after the original sale. However, these provisions don't apply to commercial auto distributors, who may still be liable.

**AV owners not off the hook completely:** New legislation will likely still require the AV owner to carry some insurance because there will be situations where he/she is still liable for damages incurred by pedestrians and occupants of the car. The AV owner may also be liable for property damage to others involved in an accident.

**Taxi/ride-hailing regulations are not suitable for AV-based TaaS:** The current set of regulations and laws that govern the taxi and ride-hailing industries are not suitable for TaaS. New AV-based transportation laws will have to accommodate TaaS providers as owners.

**AV product liability issues:** Traditional product liability principles may not be suitable for the future AV environment. If proposed AV laws don't address this, the courts may apply existing laws in similar industries such as the airplane industry. But the industries are not analogous and the application of such laws may not be fair for the emerging AV technology. In the future, especially for semi-autonomous cars (SAE level 3 or 4), personal injury lawyers may enjoin manufacturers for possible defects in autonomous software even though on the surface an incident might appear to be caused by driver error.

**Our current notion of accidents as the inevitable norm in transportation:** Autonomous cars that have been advertised as eliminating all human-error caused accidents may challenge the notion that accidents are a "frequent and inevitable contingency of normal automobile use." The AV industry has set the bar for accident-free driving much higher and courts may hold manufacturers to that expectation. Toyota has described the ultimate goal of its driverless-cars program as "creating a car that cannot be responsible for a collision." Courts may treat these types of goals as binding expectations of the ultimate users of these products.

**Cyber-attack and hacking responsibility and protection issues:** AVs will be highly computerized and connected vehicles with vehicle-to-vehicle (V2V) and vehicle-to-infrastructure (V2I) communication. They will have black-box-like recorders that will track the entire trip's information. This private information could be hacked. Worse still, hackers could get into a car's operating software and take control of the car itself! This raises serious security issues. Manufacturers may be held responsible for either of these scenarios.

**Tort law and AVs:** Personal injury layers will use tort law to claim damages as well. No changes or amendments are required.

**Distracted-driving laws**: Distracted driving is one of the top three causes of accidents right now with non-AVs. Once AVs become legal and smartphone functionality is integrated, the concept of distracted driving will change completely. These recently enacted laws will need amendments in terms of their non-applicability to AVs. Even for semi-autonomous cars, the way distraction is defined will have to change.

**AV testing and compliance laws and regulations**: While SAE and NHTSA-like bodies have defined the different levels of autonomous cars, the certification mechanism will need to be defined and regulated.

## 10.3 Regulatory Issues

AV technology is moving at a good clip. It's no longer science fiction. The public has seen the cars hopping around on public roads, and Google has been publishing data for the number of times a human driver had to intervene or take over. The data shows definite progress. We're not there yet but we *will* get there in the next decade. Adoption will be slow but it will start.

However, what is not clear is whether regulators and operators (federal, state/provincial and municipal) are ready with laws, rules and regulations. The industry is urging regulators to move quickly, saying "tell us what the ground rules are." Regulators are moving slowly, saying they're not sure how everything with AV technology will turn out. But enough is known that regulators and industry can jointly put forward an interim set of regulations.

For the past two years, the media and industry PR people have been saying that the single most important impediment to the implementation of AVs is the lack of federal and state regulations. Based on a recent press release by the NCSL (National Conference of State Legislators in the United States), I feel that American legislators have made some progress, considering that there are several unknowns about the technology. Interim regulations and executive orders are good way to start the process, which can be quite lengthy. Refinements and details will come in due course, I think.

**The Current Status of AV Regulations in the United States (as of 2018)** *(Source www.ncsl.org)*
*Note: The statistics given below are based on the information in NCSL (National Council of State Legislators) website that appears dated. I think more states have passed preliminary set of legislations – though not as comprehensive as are required.*

- Since 2012, at least 41 states and D.C. have considered AV legislation.
- By the end of 2017, majority of states in the United States had introduced legislation.
- 21 states have passed legislation, and four governors have issued executive orders related to AVs.
- In September 2017, the NHTSA released new, voluntary guidance about Automated Driving Systems. It provides guidance about federal and state roles in AV regulations.

In December 2016, Michigan became the first state in the United States to pass comprehensive self-driving car regulations. Four bills were signed into law supporting both trials and use of autonomous cars on public roads, including ride-sharing and platoons (group of vehicles travelling together).. The law says that a human driver is not required in the car. Michigan's DMV requires that AV manufacturers assume full responsibility for accidents.

Massachusetts proposed a bill that would tax self-driving cars at a rate of 2.5 cents per mile to deter these cars from circling around because they can't find parking spot. This tax would compensate the state for the loss of fuel tax. The proposed law also requires AVs to be emission-free.

To summarize the complex regulatory environment surrounding AVs:

1. Autonomous vehicles do not fit into the current regulatory framework. The federal DOT regulates the vehicles that are built in terms of safety (seat belts, air bags, etc.) but it is the state DMVs that regulate and monitor the actual operation in terms of compliance.

2. Presently, there is no uniformity of regulations for AVs across state jurisdictions in the United States. They vary from a very liberal regime in Phoenix to quite strict regulations in California where AV manufacturers must report the results of their testing (e.g., how often the human driver has had to take over from the AV software) to the DOT once a year at minimum.

3. My observation is that this was inevitable in a fast-paced and far-reaching industry like autonomous vehicles. Technology often moves faster than legislation and regulations. The states couldn't have forecast the rapid pace—they've been generally in reaction mode and have had to act "on the fly" with no pre-planning.

4. Of the regulations that *do* exist, most are only interim, designed for AV trials and testing. The current federal legislation would allow 100,000 AVs on the roads that may not meet eventual federal standards (which have yet to be defined in sufficient detail). Current federal regulations don't address the needs of state DMVs, which are responsible for vehicle operation on state roads. They also don't address the issues that will arise for TaaS and ownership of AVs by individuals who may want to operate them in other states or even across international borders. Interestingly, there was an actual AV trial between Canada (the province of Ontario) and the United States (Michigan) in 2017 with officials of both governments present.

5. Consumer watchdogs say that federal governments should have performance and safety standards in place before manufacturers are allowed to operate on public roads.

6. The US federal DOT is suggesting that federal laws and regulations should replace the patchwork of state regulations with a single unified regulatory framework. However, state representatives are saying there should be flexibility in the federal

laws and regulations to allow individual states to modify some of them based on their unique requirements, such as weather, terrain and road conditions.

7. There is a suggestion among some regulators and public-sector consulting companies that they may follow a gradual approach as to which type of AVs, for which specific usage and in which geo-fenced areas AVs will be allowed. For example, fleets in less congested areas could be allowed initially. As DMVs gain experience and feel comfortable with the results, they could start allowing individual AV ownership. This approach is not very attractive to the AV industry, which wants a more liberal regime that lets the industry figure it out so long as it also takes the risks. Of course, a few deaths involving faulty AI autopilot software would raise public wrath, and it's either regulators who would have to respond to it or elected officials, who could certainly suffer the consequences at the polls.

8. State officials say that they won't be ready by 2019 or 2020 when autonomous vehicles will be ready to drive in their state jurisdictions. It's one thing to have legislation ready, but it's far more complicated to create the systems required to implement those regulations. Not only do the regulations have to be ready, but jurisdictions also need to create detailed business processes with procedures, manuals and an IT infrastructure. This will impact almost every aspect of DMV operations—a much bigger challenge.

I worked as an architect, designer and implementer of driver and motor vehicle IT systems in Ontario, Canada's largest province. I understand how long it takes to amend automated business systems for a change of this magnitude. Business analysis of well-defined AV regulations and the design and development of upgrades to current IT systems to accommodate AVs represents a major undertaking for DMVs that could easily take two to three years after the regulations are framed.

And the big question: who's going to fund these changes—would federal regulators provide subsidies to states and provinces?

## 10.4 The Ethics Perspective

Driving is not just based on the driver's guide that DMVs hand out to beginners when they start learning how to drive. Driving assumes many undocumented standards of behavior that are expected of a reasonably prudent person – these standards are not "black and white" but shades of grey. Human drivers all have their personal notions of ethical standards. The transportation regulations and laws may not consider how human drivers exercise their ethical considerations and value system. Therefore, human drivers are not punished for making instinctive though what others may consider "ethically bad" decisions that lead to accidents resulting in personal injury to pedestrians, passengers and other drivers. An example of this could be swerving to the left beyond two yellow dividing lines and hitting a pedestrian instead of swerving to the right while trying to slow down when the car at the back hits the driver while approaching an orange light. This discretion may be acceptable to the insurance company and the legal system. However, how should a reasonable prudent "Autopilot" behave? The liberal treatment afforded to a human driver may not be available to the software programmers and designers of AI "Autopilot" system. The reason is that in the latter case, it is a conscious decision of making a choice after much thought on the part of "Autopilot" programmers.

### Classical Trolley Problem Applied to AVs

This is a classic example of ethics quoted in the literature (proposed by Phillipa Foot and Judith Thomson) to emphasize a need for ethics in designing "Autopilot" software. A trolley is ready to run over either five people or one person depending on the way that signal man (in our case "Autopilot" software) pushes the switch in the tracks. Here we are concerned about the number of people we may kill in the accident. Which way should autopilot swerve the AV?

Figure 10.4—Classical Trolley Problem in Ethics (Source Medium.com)

Ever since the notion of the self-driving vehicle came about, people have been asking the question: who will make these ethical decisions, and how will they decide what course of action to take?

Our laws are not designed to handle many of the ethical choices that autopilot software might have to make. Should AVs take the same driving exam that a human driver takes, and pass so long as they obey all the traffic laws and rules listed in the Highway Traffic Act, or should they be held to an even higher standard? Is it fair to hold manufacturers to much more rigorous standards than human drivers because they install artificial brains in the cars they sell? In a word: yes. Society rationalizes this position on the basis that human drivers can exercise judgment in a number of dynamic situations that may arise in real-life driving and cannot be tested in a 30-minute driving test. The autopilot test must be far more rigorous.

Law and ethics do not always provide the same answer. Since laws for AVs do not yet exist in detail, should we take the opportunity to inject some ethics into them? This will be a challenge for regulatory bodies like the NHTSA and the DOT. Maybe regulators, on the advice of consumer bodies and civil society, can establish basic and minimum set of ethics to be considered while programming autopilot software. These basic set of ethics might represent the prudent behavior of reasonable human drivers.

### Saving the Driver First Even at the Risk of a Pedestrian
Many established OEMs feel that it is their first and foremost duty to provide safeguards that will save the human driver first in the event of a crash. That is the rationale for seat belts and air bags—they are for the occupants of the car and not for pedestrians, who come second on the priority list. That is the position of the majority of OEMs, and that is how they have been designing their non-AV vehicles for safety so far.

### Transparency in Algorithm Design
Conventional vehicles are a mass of steel, plastic, rubber tires, glass windows, and engineering components, with some electronics and entertainment parts but with no minds of their own. Humans drive them. These vehicles have no brain and no personality. AVs are different—they have brains and personalities. We do not know how they have been designed and the average AV sales specialist probably cannot describe how AV thinks, but we know that AVs see, perceive and make decisions at speeds much faster than human beings do.

In a 2017 Stanford News article, Rob Reich, director of Stanford's McCoy Family Center of Ethics in Society, is quoted asking, "Should it be transparent how the algorithms of these cars are made? The public interest is at stake, and transparency is an important consideration to inform public debate." This might be a tough suggestion for AV manufacturers to swallow—they're not used to sharing their methods. That's where regulators and legislators will come into play. At least in the academic world, ethicists and engineers are talking. Whether AV vendors will take the cue from the academics remains to be seen.

### Questions for Ethical Debate
Let's look at some oft-posed questions that ethicists and AV manufacturers will want to consider.

1. It's an acceptable premise under the current non-AV regime that cars should be safe from the point of view of owners, drivers and passengers. The notion that manufacturers' first priority is to save the driver and passengers has been accepted by the public, buyers and regulators. Should an AV TaaS provider give preference to the safety of its passengers while they are in the car as paying passengers but not after they're let out and become pedestrians? Should the next passengers' lives suddenly be more important than the lives of the previous passengers, who are now walking toward their destination?

2. Should the AV autopilot system be designed to minimize the number of fatalities or injuries no matter what, even if it means breaking traffic rules?

3. Would the public accept the notion of designing the autopilot software to give higher priority to passengers than to jay-walking or distracted pedestrians?

4. What if swerving to miss a child who has run into the road would mean swerving toward the edge of a cliff, endangering the passengers? How should the AV choose?

These scenarios and many others like them pose moral and ethical dilemmas that manufacturers, consumers and regulators will need to address before vehicles are given full autonomy.

A 2016 study by researchers at MIT, the University of Oregon and France's Toulouse School of Economics, published in *Science* magazine (Source - Science Magazine, 2016), looked closely at consumers' feelings about such paradoxes. Interestingly, most of the 1,928 study participants indicated that they believed vehicles should be programmed to crash into something rather than run over pedestrians, even if that meant killing the vehicle's passengers. "The algorithms that control [autonomous vehicles] will need to embed moral principles guiding their decisions in situations of unavoidable harm," said the researchers.

While survey participants ethically put pedestrian lives ahead of the lives of AV passengers, they also indicated they would be reluctant to buy such a vehicle, preferring to ride in an AV that gives preference to the occupants.

**The Debate Goes On**
There has been fairly extensive debate in the media and academic circles regarding the need for ethical considerations in the design of AI-based autopilot systems. Stephen Zoepf, executive director of CARS (Center for Automotive Research at Stanford), and Chris Gerdes, Professor of Engineering at Stanford University, argue that agonizing over the trolley problem is "not productive. People make all sorts of bad decisions. But AI can and should make better ethical engineered decisions more often." For this reason, MIT has designed a platform called the Moral Machine (moralmachine.mit.edu/) to collect opinions on how AI should make ethical decisions.

On a more practical basis, we also need to figure out how fast we'll allow AVs to drive—do we restrict them to the speed limit (and should the limits change?), or do we allow operators and owners to select personal preferences for aggressive driving if road conditions allow, as human drivers do? Should there be designated lanes for AVs?

**Summary**
*In this chapter, I've described further perspectives in the AV industry. While the public expects insurance premiums to go down, the insurance industry is not saying by how much. Will OEMs take on liability in all cases? That's not clear either. Legislators and the legal profession have not yet figured out how laws should change with AVs. The regulatory regime needs to be defined in detail as well. Finally, there are many questions about ethics in this new world. Can we codify ethics? There are lots of questions and lots of discussion... but not many clear-cut answers.*

# Chapter 11

# A Methodology for Balancing Multiple AV Perspectives

*We live in a complex world. Different stakeholders of technology innovation and consumers who benefit from it often have diverging perspectives. Society at large depends on legislators and regulators to strike a balance between these viewpoints and priorities. In this chapter, I'll propose a simple methodology for striking this balance.*

In the preceding chapters, I've described various perspectives from the eyes of different stakeholders in the AV debate. Who is right? Who is wrong? I'm going to describe a methodology to balance these perspectives. But before I do that, allow me to take a little detour by recounting a fable—the famous Indian fable of six blind men who go out to discover what an elephant looks like. In fact, I'll let the English poet John Godfrey Saxe tell the story through his poem "The Blind Men and the Elephant": *(Author_Saxe, John Godfrey)*

*It was six men of Indostan*
*To learning much inclined,*
*Who went to see the Elephant*
*(Though all of them were blind),*
*That each by observation*
*Might satisfy his mind.*

*The First approach'd the Elephant,*
*And happening to fall*
*Against his broad and sturdy side,*
*At once began to bawl:*
*"God bless me! but the Elephant*
*Is very like a wall!"*

*The Second, feeling of the tusk,*
*Cried, "Ho! what have we here*
*So very round and smooth and sharp?*
*To me 'tis mighty clear,*
*This wonder of an Elephant*
*Is very like a spear!"*

*The Third approach'd the animal,*
*And happening to take*
*The squirming trunk within his hands,*
*Thus boldly up and spake:*
*"I see," quoth he, "the Elephant*
*Is very like a snake!"*

*The Fourth reached out an eager hand,*
*And felt about the knee:*
*"What most this wondrous beast is like*
*Is mighty plain," quoth he;*
*"'Tis clear enough the Elephant*
*Is very like a tree!"*

*The Fifth, who chanced to touch the ear,*
*Said- "E'en the blindest man*
*Can tell what this resembles most;*
*Deny the fact who can:*
*This marvel of an Elephant*
*Is very like a fan!"*

*The Sixth no sooner had begun*
*About the beast to grope,*
*Then, seizing on the swinging tail*
*That fell within his scope,*
*"I see," quoth he, "the Elephant*
*Is very like a rope!"*

*And so these men of Indostan*
*Disputed loud and long,*
*Each in his own opinion*
*Exceeding stiff and strong,*
***Though each was partly in the right,***
***And all were in the wrong!***

I apologize if any AV stakeholders are offended by the suggestion that they might be in the wrong. I simply mean to suggest that the holistic perspective of society at large is different from that of individual groups and that it's important to look at the whole picture. There are two questions we need to ask. The first is from whose perspective should we decide on the optimal way forward? The second is how do we arrive at a holistic or optimal perspective? AVs should be predominantly designed and built to satisfy consumers' needs. Consumers, of course, belong to multiple generations—baby boomers, Generation X, millennials (Gen Y) and Generation Z; if they have a mishmash of different opinions, so be it. But they're not the only group who should get a say. As a composite society, we must satisfy multiple demands. I suggest that the holistic perspective should reflect a composite of the viewpoints of consumers, suppliers and other stakeholders. Nobody should lose—everyone should win, to a certain degree, at least.

Now I want to propose an approach that might help us arrive at that holistic perspective. Let's define a holistic perspective or an optimal approach toward solving a difficult problem as one that is not preconceived and that takes into account all perspectives but assigns more weight to more important perspectives.

Also implicit in this discussion is the fact that we need regulators and legislators to influence the overall direction of this optimal AV development.

### The Science of Arbitration across Competing Viewpoints

While I was at the University of Toronto, I did my Master's degree in management science. One of the things I studied was mathematical techniques of linear programming that can be used if, in a mathematical model of a multi-variant problem, the objective function of the variables is deterministic. If the objective function is not deterministic, but is instead *probabilistic* (i.e., it can vary depending on other factors), a more sophisticated technique called stochastic programming can be used. These techniques allow us to develop an optimal proposed solution when there are a number of competing variables with different objective functions. It would obviously be an extremely difficult exercise to build a mathematical model for the question of AVs and determine an objective function of the different stakeholder perspectives in the AV conversation. (It would make a good Master's project for a management science student!) This book is not an academic research project, so I will give up on that solution right here.

Instead I'll propose a simpler process of weighted-criteria decision-making. I've used this method in my consulting practice and it's commonly used in business. In this process, we assign weights to the different perspectives (criteria) that different groups have on an issue, and the total weight that we assign to all perspectives is 100. The theory is that that the group that has maximum impact on a particular decision should be given the highest weight. Others are given proportionally lower weights depending on their importance and influence on the objective that we are evaluating. Ideally the weights are assigned by consensus among a group of un-biased experts (or, since everybody has some bias, let's say the group should have *minimum* bias).

Once the weightings are assigned, a group of experts drawn from each group in proportional numbers then scores the group's perspective on each specific strategic direction that we are trying to evaluate. The score is out of 100. Then we add up the scores for all the groups. If the score is closer to 100, then the recommendation is to follow that specific strategic direction. If, on the other hand, the score is closer to 0, then the recommendation goes against the strategic direction.

## Assigning Weights to Stakeholders

In Table 11.1, I've proposed weights to be assigned to the different stakeholder groups' perspectives, and I explain my rationale following the table.

|   | Groups with Perspectives on AV Deployment | Characteristics of the Group | Suggested Weight |
|---|---|---|---|
| 1 | Potential customers: Baby boomers and traditionalists | Comprise 23+10% of the US population Important customer group with financial clout and influence over younger generations Very keen on safety but less eager to buy fully autonomous vehicles | 12% |
| 2 | Potential customers: Generation X plus millennials (Gen Y) | Comprise 44% of US population Most attractive group to buy or adopt AVs | 18% |
| 3 | Potential consumers: Generation Z | Comprise 23% of US Population Technologically savvy Majority will adopt AVs or TaaS | 10% |
| 4 | Established auto OEMs | Important group among potential AV suppliers Have the manufacturing assembly lines and sales/service network. Acquiring the know-how to build AVs Are pragmatic business organizations Consumer-centric | 20% |
| 5 | Newcomer AV vendors | Leaders who have the AV R&D skills (AI, sensors) Lack assembly-line capacity | 10% |
| 6 | Academia researchers, innovators and start-ups | Key players in creating the technology basis Since their main pursuit is creating the technology, they have less understanding of how to commercialize it—which will make or break any disruptive technology | 5% |
| 7 | Insurance companies | Will be seriously impacted (revenues will go down) Have small contribution and influence | 3% |
| 8 | Road infrastructure owners: city transportation departments | Very important group in creating environment for accelerating AV adoption Minimal budget for improving infrastructure | 14% |
| 9 | Regulation monitors: DMV and public safety agencies (e.g., police) | Have to operate and monitor the regulations Need to upgrade their processes and retrain their staff | 8% |
| 10. | Regulators and legislators: target decision makers for whom this model is built | We are building this model to allow this group to create regulations and draft legislation | 0% |
|   | Total |   | 100 |

Table 11.1: Proposed weights to be assigned to each stakeholder group

## Rationale for My Weights

I recommend that regulators form a focus group drawn from all the relevant subgroups that could come to a consensus on the weights. I am providing a rationale (biased with my personal opinions and bias) for assigning these weights is as follows:

### Consumers

Within the consumer group, each of the generational groups has a varying amount of influence on AV adoption. Baby boomers constitute 23% of the total population in the United States, with traditionalists (those older than baby boomers) constituting another 10%. They also have the financial clout to influence AV purchasing decisions over the next ten years. Therefore, for the short and medium term (up to ten years out), they will have considerable influence over the industry and thus should be given a fairly good weight; I propose a weight of 12% for this group.

The most important group consists of Generation X and Y, representing about 44% of the US population. This group is the prime target for the AV market. Therefore, this group should have maximum say in the development and introduction of AVs. Hence I'm suggesting a weight of 18% to this group. Generation Z should carry a weight of 10%.

### Suppliers: Eliminating Unfair Advantage due to Lobbyists

In the absence of my approach, regulators and legislators are often unduly influenced by the lobbyists, trade associations and PR firms who work for the suppliers. In my process, the focus group will try to build a consensus on weights after an open debate and discussion, and the regulatory body that oversees the process will be bound to honor those weights. That will eliminate the influence of lobbyists from the conversation. I suggest that suppliers (established OEMs and newcomer AV vendors) should be given a total weight of 30%., divided as 20% for OEMs and 10% for technology challengers.

### Infrastructure Owners and Regulation Monitors Must Have an Important Say

Federal regulators like the NHTSA and the DOT in the United States and similar bodies in other countries set the policies and regulations for AVs on interstate highways, and state DMVs administer transportation policy in each state. However, the real management of road infrastructure (construction, maintenance and upgrading) is the responsibility of cities' traffic/transportation departments, so I suggest they be given a weight of 15%. Similarly, public safety agencies monitor the traffic laws and regulations. Both of these groups are important stakeholders in the AV discussion. They must be prepared to accept the challenge that AVs will bring about. I'm recommending a weight of 14% to the infrastructure owners and 8% to traffic monitors i.e. the traffic police.

### Academic Researchers and Insurance Companies

While academia is very important for the discoveries they make and have an independent viewpoint, they have a smaller role in commercialization of the technology. Nonetheless, they are being assigned a weight of 10% because of their independence and advisory role in society.

Let's try this methodology out for the first question out of several strategic questions that are being debated among AV industry stakeholders. Here is a sample of these questions:

**Q1: What level of regulations should we have for AVs? Should we have strict regulations to ensure a phased and successful introduction or should we let the industry and consumers fight it out in the market?**

Q2: Should we allow the steering wheel and pedals (acceleration and braking) to be removed from SAE level-5 AVs i.e. should we let the autopilot handle everything?

Q3: Should AV suppliers be allowed to restrict AVs to TaaS providers only and not sell these to private buyers?

Q4: Is incremental approach to AV introduction better than going for SAE level 5 directly?

Q5: Should we reserve lanes for AVs?

**Question1: Level of Regulations—Strict or Loose?**
Consumers would prefer tight regulations in the beginning, to be relaxed as they gain confidence about all the industry's claims of safety. The public would also like regulators through the DOT and the DMVs to establish independent testing and certification facilities that will ensure that the industry is delivering products as advertised. Baby boomers along with Generations X and Y want vendors to have clearly defined sales contracts that accept liability for accidents. Just like in the telecom industry, regulators must protect consumers against unjustified prices for OTA software updates that don't provide additional functionality. The supplier community, on the other hand, will want the least amount of controls so that *they* can decide what's in customers' best interests. Infrastructure owners will want strict regulations and may ask the industry to help fund upgrades to the infrastructure to accelerate the adoption of AVs. If industry claims of reduction in injury and death are supported by real-life statistics in the first five years, infrastructure owners may support some relaxation of regulations.

Let's tally the score in Table 11.2 and see which way the wind blows. With my own bias, I have suggested the scores of various groups – as to how they will score. I am suggesting that baby boomers will give a score of 100 out of 100 implying that they are in favour of firm regulations. On the other hand, established OEMs will give a score of 25 only out of 100 implying that they would like very light regulations and let the industry have most of control.

| | Question 1 - Level of Regulations—Strict or Loose | | | |
|---|---|---|---|---|
| | Groups with Perspectives on AV Deployment | Suggested Weight in % | Score Out of 100 | Weighted Score (Score*Weight/100) |
| 1 | Potential customers: Baby boomers and traditionalists | 12% | 100 | 12 |
| 2 | Potential customers: Generation X plus millennials (Gen Y) | 18% | 80 | 14.4 |
| 3 | Potential consumers: Generation Z | 10% | 60 | 6 |
| 4 | Established auto OEMs | 18% | 25 | 4 |
| 5 | Newcomer AV vendors | 7% | 30 | 2.1 |
| 6 | Academia researchers, innovators and start-ups | 10% | 80 | 8 |
| 7 | Insurance companies | 3% | 100 | 3 |
| 8 | Road infrastructure owners: city transportation departments | 14% | 100 | 14 |
| 9 | Regulation monitors: DMV and public safety agencies (e.g., police) | 8% | 100 | 8 |
| 10. | *Regulators and legislators: target decision makers for this model* | 0% | 0 | 0 |
| | Total | 100 | | 71.5 % |
| | "Balanced" recommendation from all the stakeholders: | Go for fairly tight regulations initially (for the first five years). Relax regulations later, if AV implementation is successful. | | |

*Table 11.2: Weighted decision-making for strict vs. loose regulations*

We take the score of a group and multiply it by its weight. Then we add up the weighted scores for all the groups. The final total of weighted scores determines the course of action to be taken. In this example, a total weighted score of 71.5% suggests a fairly tight set of regulations – not iron clad but pretty tight.

You can see how this methodology can be used to make recommendations for difficult questions where different stakeholders have different perspectives. All I am saying is decisions should be broad-based and influenced more by the factors or groups that are most important. Will the regulators use this methodology? I am not sure. I can only hope.

**Summary**
*Decision-making in the business world is not an exact science. However, the discipline of management science often provides help in making it a soft science. Here I have given a relatively simple weighted-criteria based methodology for making difficult decisions for multi-variant business issues. I hope the AV industry is listening.*

# Chapter 12

# Challenges Facing the AV Plus Industry

*It is no surprise to me that an emerging technology like AV Plus, which will completely disrupt the way we transport people and goods, is facing many challenges. It happened during the industrial revolution. Grudgingly, patiently and out of necessity the business world will adapt and change. It's simply a matter of time.*

Any new technology faces significant challenges for adoption, if it disrupts an existing mode of serving the needs of a population. The introduction of AVs is not like the introduction of smartphones in 2004 because there *was* no established smartphone industry or large customer base before their introduction. Smartphones met no resistance from users. They were received enthusiastically. AVs, on the other hand, face many challenges because they are disrupting a hundred-year-old industry. Also, they're expected to change how we achieve one of our most basic needs—moving from one place to another. Movement is life. Let's look at some of the most significant challenges that the industry faces.

## 12.1 Maturity of AV Plus Technology

The quest for self-driving cars began more than 90 years ago in New York City. Today, it is no longer a dream or science fiction. But we are not there yet. If AV technology has a lifespan of 100 years, then we are in the first decade of a century in our new AV-based lifestyle. The fact is that AV technology is still quite immature and will remain so for the foreseeable future. In the next five years, a few daring souls may ride in AVs when suppliers put something out for customers to try (as in TaaS) or buy, but the masses will watch and wait for a more convincing outcome before they vote for the AV with their pocketbooks.

The components (especially sensors) that the industry is using in trial vehicles are expensive, and we still haven't got the configuration quite right. Better, cheaper and more functional sensors are being developed. The heart is pumping but the blood isn't flowing freely; it's not that there's fat clogging the arteries, they're just not wide and strong enough for all the blood we must pump to pass through. We can stammer but we can't yet talk in a free-flowing language that our listeners will understand with 100% fidelity. We can *kind of* see our surroundings but the view is still foggy and hazy, as if we're in a snow storm. If the sun is shining, the roads are well-marked and not many vehicles are clogging the roadways, AVs do well. Ironically, AVs' early functioning is the opposite of how beginner drivers' permits typically work, where new drivers are allowed to drive on city roads but not on highways—AVs can drive on highways but not in the city.

**And what about the brain of the AVs?** How mature is the artificial intelligence discipline—not in terms of its age (because it's been around for quite some time) but in terms of its efforts to mimic the human while driving? The trick is not just in optimizing the *seeing* function, because the sensors do a pretty good job of seeing. Seeing physically (sensors picking up pixels of images or distances of obstruction) is useless if the signals the sensors and cameras detect cannot be translated into accurate perception of the physical world—that's what human nerves and brains do. How good is AI's learning ability? Is it a fast learner? Are the "teachers" guiding this learning in a trial-and-error-type of experimentation with all the data they have? Here are a few realistic opinions from experts published in a 2017 *New York Times* article and other publications:[31]

- Even Geoffrey Hinton, the godfather of AI who leads the AV-related initiatives at the Google Brain lab in Toronto, acknowledges the need to go beyond neural networks to expand the potential of AI in improving perception and developing higher, softer forms of intelligence that more closely mimic the human thinking involved in driving. We, humans, have our weaknesses but we have a lot of higher-level strengths as well.
- "Eric Horvitz, who oversees much of the AI work at Microsoft, argues that neural networks and related techniques are small advances compared with technologies that would arrive in future. 'Right now, what we are doing is not a science but a kind of alchemy,' he said."
- "Oren Etzioni, CEO of the Allen Institute for Artificial Intelligence, based in Seattle, lamented what he called the industry's myopia. Its current focus on neural networks, he said, will hurt the progress of AI in the long run."

While we have taken huge steps in using deep learning for autonomous driving, this is just the beginning of our journey. We have a long way to go before we can say, "we have arrived."

## 12.2 Complexity of the AV Technology—It is a Moonshot

The introduction of fully autonomous vehicles is a dramatically complex undertaking. Some experts compare it to the space project lunched by President Kennedy during1960s, it is a moon shot in that it may require the entire nation to be fully behind it. In other words, it needs to be a coordinated, industry-wide effort with a single-minded goal. Here are some of the reasons it's such a complex effort.

- The target is beyond our current state-of-the-art computing capability. It's never been done before, so it's hard to know when we'll get there.
- There's no master plan for the industry as whole. There's nobody in the White House or at any international business forum creating a national or international plan. In fact, how *can* we have a master plan in a highly competitive environment? Nobody wants to give away their intellectual property.
- Even though some of the brightest scientific minds around the world are working on AVs, they concede that it's complex and that new ground will have to be broken.
- With 60-plus computers controlling the different functions in the AV Plus vehicle, there's no single operating system that has proven itself capable of managing them all. Established OEMs have no experience building an operating system,

and newcomers, who *do* know how to build complex operating systems, don't know how to build a car within a specified timeline.
- It's been estimated that AV software code may contain as many as 300 million lines—compare that with the Boeing Dreamliner, which has only 6.5 million lines of code and modern car has 100 million lines of code (see Figure 12.1) [32][33]. Software reliability experts suggest that the likelihood of error increases exponentially with additional lines of code. The code will most certainly have its failures. If it fails at a critical juncture, it could lead to a fatal crash.

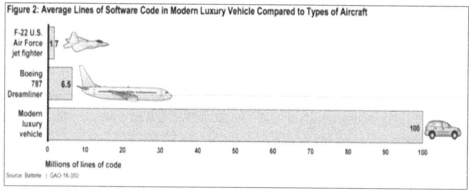

*Figure 12.1: Lines of software code in airliners vs. modern luxury vehicles (Source: US Government Accountability Office GAO-16-350)*

- We accept that most car accidents are due to human error and that a human driver is not as reliable as an AV. However, the future AV, with all that software code and the great number of computers installed, will never be 100% reliable. Auto components fail, computing hardware fails, software fails to handle unforeseen situations (which is not always the programmer's fault—the unforeseen situation may not have been included in the specifications in the first place). When an AV fails, it can be dangerous or even fatal, and we're going to have to come to grips with some risk.

## 12.3 AVs being Mission Critical Require Redundancy Design

Failure of AVs can lead to life and death situations. Therefore, all critical components must have a built-in redundancy and fail-safe mechanism. Remember, AV occupants will generally have very little knowledge of how the vehicle works and, except for accessing a remote helpline, will be incapable of assisting an AV in an emergency situation. If a critical component like the AI server fails, could crash into another vehicle. If this redundant design is implemented afterwards in a subsequent version of the level 5 AV, then entire OS has to be redesigned and re-coded. What a difficult task that could be.

If you're familiar with computer OS design and tandem fault-tolerant systems for financial transaction processing (used in banking systems and stock exchanges), you'll appreciate the significance of this issue.

## 12.4 AVs Cannot See as Clearly as the Human Eye

Sensors and cameras installed in an AV may have a 360-degree view of the surroundings but even after assembling all the inputs through sensor fusion into a comprehensive view, an AV's perception of its surroundings is still not as good as what can be detected by the human eye supported by the human brain. We marvel at what AI can do, but let there be no mistake: the human brain is a computer that has not been replicated so far.

Although I'm a computer professional with knowledge of IT as a discipline with all its strengths and weaknesses, I'm not an AI expert. However, it's my understanding that the current state of the art in AI is in trying to teach an AV's AI software what a dog looks like and what a pedestrian looks like. A human child knows that practically when he or she is born. AI still has a lot of learning to do till it is able to mimic a human driver in all situations.

Geoffrey Hinton himself, as reported in the same *New York Times* article quoted earlier, "points out that his [neural network] idea has had its limits. If a neural network is trained on images that show a coffee cup only from a side, for example, it is unlikely to recognize a coffee cup turned upside down." That would certainly be an issue for AVs. The article continues: "Now Mr. Hinton and Sara Sabour, a young Google researcher, are exploring an alternative mathematical technique that he calls a capsule network. The idea is to build a system that sees more like a human. If a neural network sees the world in two dimensions, a capsule network can see it in three."

*Figure 12.2: Professor Hinton and Sara Sabour of Google (Source: New York Times)*

## 12.5 The Handoff Conundrum[34] (Source - CARS Stanford CA, 2016)

There is a major debate on the topic of AI-to-driver handoff in level-3 and level-4 AVs, in academic circles, industrial development labs and business strategy departments of AV stakeholders. It is truly a conundrum because there appears to be no perfect solution at this stage of AV development.

Here's why. Let's assume that our target is SAE level 4 (an AV capable of driving under all conditions but that supports human-driver control) or level 5 (an AV in which no human-driver intervention is supported). We know that there are two camps in the suppliers' corner. In the first camp, we have Waymo and a few others in the later phases of the AV evolution who support a direct route to level 5; in the second camp, we have others who feel that we should leave the ultimate control with the human driver, for the time being.

The first camp feels that we should forget about all AV levels except level 5, because human drivers make for poor backups. Proponents have good reasons for their position. Both CARS and MIT's CSAIL (Computer Science & Artificial Intelligence Lab) and AgeLab have researched the subject and essentially agree that we cannot depend on a human driver to take over in an emergency. The human driver may not be fully or may have dosed off to sleep. Reaction time may be too long. And human driver may not know what to do either because he/she has not been driving in difficult situations.

Holly Russell, a former graduate student at Dynamic Design Lab at Stanford, published a paper on her research showing how human behavior and motor activity changes if the driver has not been participating actively in driving in the immediate past. She tested the reactive skills of 22 drivers on a track and found that the transition to successful takeover of the vehicle could be rough. The issue is attention and situational awareness. The situation may require a split-second response but the human driver might take a few seconds to collect themselves, figure out what to do and take effective action.

In yet another case (described in a _Wired_ article[35] in January 2017), Erik Coelingh, head of Volvo's safety and driver-assist technologies, changed his approach toward AV design after he participated in a simulator run of an experimental AV and, while he was busy playing a game of Dots on his iPad, was asked to take over. The alert came when he was about to earn a high score.

> He had no idea of what was happening on the "road" or how to handle it. "I just realized," he says, "it's not so easy to put the game away."
>
> The experience helped confirm a thesis Coelingh and Volvo had been testing: A car with any level of autonomy that relies upon a human to save the day in an emergency poses almost insurmountable engineering, design, and safety challenges, simply because humans are for the most part horrible backups.
>
> And so Volvo, and a growing number of automakers, are taking you out of the equation entirely. Instead of developing autonomous vehicles that do their thing under most circumstances but rely upon you take the wheel in an emergency—something regulators call Level 3 autonomous capability—they're going straight to full autonomy where you're simply along [for] the ride. (Source - Wired Magazine, 2017)

To ensure that level-3 and level-4 AVs work effectively and safely, the AV gadgetry must ensure that the driver is fully alert—has not dozed off and is not engrossed in some activity like watching a movie (like the Tesla driver who crashed in Florida). To ensure the driver's alertness, vendors must install cameras and sensors inside the vehicle to watch them. This poses a privacy concern. Vendors must also use visual, audio and heptic alerts to get the driver's attention. Most of all, we need to make sure that the AV brain can predict emergency situations let's say 8–10 seconds ahead—the time it could take for a human driver to take over. Can the AV brain think that far in advance? The current generation of AV brains can't.

**That is why it's a conundrum.** Solving the takeover problem is just as difficult as developing foolproof SAE level-5 software. It is for this reason that Waymo decided to go for the target environment of full autonomy – SAE level 5.

## 12.6 Severe Weather Conditions[36] *(Source: Detroit Free Press)*

It is quite understandable that the majority of AV trials are being done in relatively good weather conditions on the west coast. We wouldn't want AV prototypes to be put to the test under severe weather conditions on day one. Yet at some point they must be ready to grow up and handle snow storms, rain storms and high winds. It's precisely for this reason that the American Center of Mobility has opened an AV testing facility in Michigan. Inclement weather conditions may make various sensors inoperable or ineffective. I expect that in level-3 and level-4 AVs, if the sensors are inoperable or not getting enough reliable data or if the AV is skidding, the vehicle will turn the control over to the backup human driver. How the robo-drivers of various AV manufacturers will deal with these situations in level-5 vehicles is not clear, at this stage. Many more tests will be conducted—the Canadian provinces like Ontario and Alberta with severe winter conditions will be involved in future tests.

## 12.7 Beyond the DMV Handbook—Unpredictable Human Drivers, Pedestrians and Cyclists

The DMV handbook spells out the basic rules of the road. Real-life driving skills go well beyond what's in the handbook, requiring a human driver to use their judgment in a wide variety of undocumented situations. Robo-driven AVs are expected to be safe drivers and good citizens of the road, following all the traffic rules, aware of everything around them and giving way to pedestrians even if they are jay-walking. However, we do not know how human drivers are going to react to AVs that do not have a human being sitting in the front. Some human drivers may try to take advantage of the situation and overtake AVs even if they don't have the right of way (a case of "non-humanoid" discrimination!). This could cause dangerous situations. How auto companies design their AI software to deal with situations like this will determine whether AVs continue to face such dangers.

Some cities have been reserving bike lanes and giving them preferred access for several years now. Could robo-autopilot co-exist with bike users in an amicable fashion?

There are many other driving situations where the robo-driver needs to be programed correctly. Often when someone wants to make a left-hand turn across constant oncoming traffic, the driver may wave the oncoming driver who may slow down, letting the driver make the left—we need to determine how an AV should behave in such a situation. Similar situation happens when you want to change lanes in a fast-moving traffic. Still another situation to account for is how an AV will behave when a human (a traffic cop, a crossing guard, etc.) is directing traffic due to a broken traffic light, an accident, a slow zone around a school or construction site or whatever else might call for it.

Some AI companies, like Drive.ai, are attempting to tackle these less frequent situations and are creating AI algorithmic modules to handle them. They're experimenting with early versions of AI's conversational interaction with other robo-drivers in the vicinity or with human drivers to try to come to some "civic sense savvy" agreement that will determine who will cede the way to whom. This is still a work in progress and the hope is that the user community will come to an acceptable understanding that the industry can adopt as a de facto standard. (Source drive.ai, 2017)[37]

Figure 12.3: Interaction with pedestrians and other drivers (Source: drive.ai)

## 12.8 Connected AV Issues—Standards & Implementation

An AV Plus car of the future will be safer, smarter, autonomous, greener and connected—all in one. That's my definition of Plus. The network application architects are exploring lots of ideas for what they can do with the connectivity of the car; it'll be useful in many ways.

First and foremost is cellular connectivity—having our smartphones connected to the cellular network through Bluetooth for voice calls and data applications. That's where the Apple CarPlay and Android Auto interface support becomes a requirement. This part is easy.

The second use is for OTA software updates and fixes. What we don't know is who will pay for such updates if they're done on a cellular network—the manufacturer or the owner?

The third application of connectivity is Wi-Fi within the car to make the AV Plus an infotainment hub. Frazzled parents will not only be able to stream Netflix or YouTube to keep their kids entertained, but they'll be able to relax and enjoy the entertainment themselves as AV occupants.

The fourth connected-car application, and arguably the most important for a self-driving car that's aware of and is in touch with its surroundings, is for V2V (vehicle-to-vehicle) and V2I (vehicle-to-infrastructure) communication.

### Understanding V2V, V2I and V2X

It is anticipated that the autonomous vehicle communications architecture will involve some form of communication with other vehicles in its vicinity: V2V. This may be used for exchanging not only location and speed information but also information about each car's intent to make a left or right turn or other driving maneuvers.

V2I communication is between the vehicle and the infrastructure such as street lights, electronic road signs, emergency-response vehicles, etc., all of which will need to be upgraded to handle such communication.

V2X is a general term for communication between an AV and the moving parts of a transportation system. V2X communication includes V2V and V2I. Some vendors are developing proprietary products for this communication and some auto manufacturers have started installing early and proprietary versions in their luxury brands.

### The DSRC vs. 5G Debate

Industry network experts are considering specialized communications protocols for AV communication. Two protocols are being explored so far—5G and DSRC (Dedicated Short Range Communication).

Proponents of both standards are saying that their standard can meet the needs of AVs. As a seasoned network professional, I feel that these arguments are not based on engineering evaluation of the standard to meet latency, speed, capacity and cost points of view. I have yet to see an objective assessment of the two that takes into account the following important factors:

a) Has the protocol become an international standard, adopted by international communications bodies? If not, how far along is it in this process?
b) Considering the inherent delays in 5G implementation for massive usage, will the network service providers be ready for AV connectivity?
c) What is the network capacity of the protocol for the intended purpose considering the spectrum allocation licensed for it? Is there sufficient capacity for such massive amounts of data to be transmitted in a congested intersection?
d) What are the maximum upload and download speeds the network can support?
e) How much would it cost to transmit data from a single AV? Who will pay for this data?
f) Has any network modeling been done to determine the latency in transmitting critical V2V data at a congested intersection ?
g) How are we going to secure these networks from cyber-attacks?

h) Which AV data should be stored on the AV and which data should be in the cloud from performance and privacy points of view?

Proponents of DSRC argue that it can meet all the requirements for V2V and V2I. DSRC hardware modules are available from NXP in the $100–$200 price range. OEMs like GM started installing these modules in luxury cars (such as the Cadillac) in 2017. Regulators in Europe and the United States seem to have blessed DSRC to some extent. It does function in bad weather, and it can support high speeds and low millisecond latency. But it only works in a short range, unlike 5G, which works across a wide area. DSRC is also not suitable for infotainment applications, which 5G *can* easily support, albeit likely carrying quite a high cost.

In my view, both DSRC and 5G could be used for specific AV Plus applications. A lot of analysis has to be done before the industry decides which application is best suited for which network. In Table 12.1 I've given a brief outline of the major features of each one.

| | Feature | DSRC | 5G |
|---|---|---|---|
| 1 | Major technical specs | Short to medium range Communication mechanism for ITS (Intelligent Transportation System)—used for toll collection, etc.<br><br>Ideal for short messages like SMS<br><br>Low latency<br><br>Highly secure and suitable for V2V because spectrum is dedicated | Target is higher data rates (100 Mbps metro and 1Gbps for single floor applications)<br><br>Superior spectral efficiency<br><br>Major application is the wide-area wireless cloud |
| 2 | Position in product lifecycle | Early in its life but ahead of 5G | Standard just approved by 3GPP<br><br>Rapid progress in trials<br><br>Early in concept—final specs post 2020<br><br>Implementation after 2020<br><br>Verizon pilot in 2018, commercialization in 2019–2020. |
| 3 | Most suitable application | Short messaging over short distances | High-speed, wide-area communication for updating maps, road signs from remote server<br><br>Machine to machine (IOT) for getting sensor info etc. |

| | Feature | DSRC | 5G |
|---|---|---|---|
| 4 | Spectrum: licensed or not | FCC: 75 MHz in 5.9 GHz band<br><br>DOT wants exclusive use for transportation, but FCC has yet to approve it<br><br>Europe: only 30 MHz allocated | Various bands in the 1–6 GHz range<br><br>(mostly 3,300–4,200 MHz and 4,400–4,990 MHz)<br><br>Licensed band, approval not confirmed |
| 5 | Capacity | Limited | Very high for cloud applications |
| 6 | Hardware availability | Yes, some early proprietary versions available (e.g., NXP) | Early R&D samples based on proprietary version being piloted (from Qualcomm) |
| 7 | V2V | Yes, supported | Not suitable (in my opinion) |
| 8 | V2I | Yes, supported | Not suitable (in my opinion) |
| 9 | V2X | Maybe—depending on distance | Maybe |

Table 12.1: DSRC and 5G comparison for AV applications

## 12.9 No Unified HMI Standard(s)

When we climb into the driver's seat of a new car, most of us feel immediately comfortable that we can drive it without needing a major orientation to the car's design. The steering wheel is a standard dimension, the brakes are within an inch or two of where they were in our old car. The accelerator is on the right side of the brakes. The console is more or less standard, and the buttons are all marked. "Auto" on the headlight control means the same thing on all cars and the windshield washer is activated more or less the same way. Even the seat adjustment is very similar on most cars.

Now look at the driver-assist and automation (e.g., ADAS) features that have started being installed in luxury cars today and with which I'll share some of my personal experiences in my 2017 Mercedes C300 with Intelligent Drive. From one manufacturer to the next, everything is different. The user interface is even different between the Mercedes E-class and S-class! The navigation systems work differently on all cars and the interfaces use different toggles and cursor-moving mechanisms. Some vendors use a touch screen—others use a rotating toggle (which, in my opinion, makes it very hard to move from A to Z on a linear alphabet keyboard. It used to take me 10 seconds to enter my destination on my cheap "TomTom" GPS system's touch keyboard; in my Mercedes, it takes easily 30 seconds to enter the same address because it doesn't support touch screen). Voice interfaces work reasonably well for somebody who grew up or went to school in the United States or Canada, but they aren't very forgiving of voices with even mild accents. I have lived in Toronto for 50 years; I have a bit of an accent but others rarely ask me to repeat what I say. Alexa has no problem with my commands. And yet I to tried train my Mercedes and it didn't learn very well at all. Such is the state of NLP (natural language processing) as implemented today. I know it's getting better.

How about parallel parking? (Letting the Mercedes handle it is not easy for this faint-hearted older driver—it's essentially a question of trust.) The interface is different from Lexus to Mercedes. Younger generations may accept minor variations across brands but older users generally want interfaces close to what they're used to. Intelligent dynamic cruise also works differently on different brands. Each OEM is trying to brand their HMI as the best, but it's very hard to say at this point which HMI features will stand the test of time.

Now, let's consider a truly autonomous car in 2021 or 2025 (you pick the year). You are the lucky buyer of a level-4 AV from Mercedes or the Waymo/FCA team. You'll probably spend a day in a Mercedes classroom learning how to use the interface, then you'll get to practice using the interface by driving your car, through lots of trial and error. But what if you switch to Tesla the next year? Remember you are not going to keep AVs for ten years. Will you have to learn yet another user interface? "It'll be quite intuitive," the Tesla salesperson will tell you, but will it? The lack of a common user interface will slow adoption, I think.

The reality is that there's nobody leading the HMI charge. The auto industry is not like the smartphone industry where there are essentially two interfaces—iPhone and Google's Android. Can we hope that there will be only 2–3 HMI standards in an industry where there are 15–20 large OEMs? I hope so. I don't think OEMs should use their HMI platforms as branding opportunities—the core HMI should work in the same way across the industry.

## 12.10 Cybersecurity and Hacking Issues

In the brave new world of AVs, we will be dependent on AVs with 60+ computers connected to another computer-based infrastructure. Every one of our movements will be recorded somewhere. This new world will require extremely tight cybersecurity and protection from criminals and terrorists. There's a clear opportunity for bad elements to break into the system. The US DOT requires that AVs be capable of being controlled by a remote operator. Imagine what would happen if a criminal got control of the network? You might not reach the destination of your choice but the destination of a *criminal's* choice instead, if you reached one at all! This could be a very dangerous situation. The US Government's Accountability Office (GAO) has highlighted in a report to the US legislators[38] some of the implications and steps that need to be taken. Can we expect the industry to take care of this without strict regulations?

## 12.11 Evolving Regulations for a Hybrid Environment

For quite some time, there will be a hybrid environment wherein there will be very few AVs driven by robo-pilots and lots of non-AVs driven by human drivers sharing the road. Over time the breakdown will shift, but what's not known is at what rate the adoption will happen and how long it will take to reach the 50/50 inflection point. There are a few very optimistic projections (I call them the "everything going in their favor" guesstimates by AV promoters). I'll analyze some of these projections in Chapter 14.

## 12.12 City Infrastructure Not Ready for AVs

Human drivers are not always attentive, are distracted from the driving task and are not always able to make the best decision in a given situation. But human drivers can drive in many difficult road environments. They can drive in the congested streets of New Delhi where there are no lane markings or sometimes three lanes of cars driving in two marked lanes. AVs do well in, for example, the clearly marked lanes of a newly developed subdivision in Austin, Texas. The state of city infrastructure in terms of road markings, stop signs, road signs and street lights varies from city to city, state to state and country to country.

Cities would love to upgrade their infrastructure to make every street AV-ready, but finding the funds for that is another matter entirely. City budgets are already under a lot of strain and many governments are cutting subsidies and funds for these projects. Will each city need AV inspectors who drive around and certify AV-worthy streets?

## 12.13 AV Business Models Are Speculative and Optimistic

Promoters of AVs have made the following very optimistic predictions:

a) The auto industry will soon start delivering AVs that are fully self-driving. These AVs will drive almost flawlessly and will make no mistakes on the road. Traffic will move smoothly. As a result, there will be very few accidents, and casualties on the road will be reduced by 90%. Since AVs will be safer than their old cars, the public should and will flock to these AVs.

b) There will be fewer cars on the road because the TaaS model will rapidly replace private ownership —thus reducing congestion. Therefore our current road infrastructure will suffice.

c) Under the TaaS model, most AVs will be on the road most of the time driving somebody from point A to point B, freeing up parking spaces.

Throughout this book I've outlined why I have difficulty trusting these arguments. Switching from personal ownership and private driving habits to a shared drive with other passengers in a TaaS AV has a big question mark.

## 12.14 Marketing Challenges

Most consumer surveys conducted by market research and/or consulting firms indicate that the general public, with the exception of the technology-savvy and the younger generations, still does not have a lot of trust in AVs. The most comprehensive survey I have come across is the one conducted by Deloitte, which surveyed 22,000 consumers in 17 countries, which showed that consumers are apprehensive and confused about how the AV revolution will happen. Auto companies like Waymo and others have started "touch, feel and try" trials that will help in building the public's trust, but much more effort is required—this may be the single biggest deterrent of faster adoption.

Another factor that will affect adoption rates is the likely pricing of AVs. Right now the bill of material for the components that go into an AV is high and R&D in AI is very expensive. It is a common business strategy to introduce a new product to the market with a high price tag. In Chapter 9, I showed that there is a huge gap between what the car companies may want to charge as price differential for AVs (around $10,000) and what consumers may be willing to pay ($1,500).

Then there's the obvious difference in the financial capacity of the different people who may want to buy AVs. Baby boomers and Generation X may have the financial capacity to buy AVs but they're less interested. Generations Y and Z, who are better targets and may want AVs, are less likely to have the financial capacity. Also, as far as TaaS is concerned, the business model assumes consumers will be willing to share their rides with other passengers. I'm not sure that consumers are ready to give up their privacy in that way. Without sharing the AVs, the TaaS cost model is flawed.

## 12.15 No Single Vendor Holds All the Marbles to Deliver AV Plus

When we review the auto vendor landscape, we see two distinct groups—the first one includes all the established OEMs, and the second group includes technology vendors from Silicon Valley and technology start-ups. There are distinct differences between the two groups.

The first group has the manufacturing savvy, supply chain discipline and supporting infrastructure to build, sell and service millions of cars. They make low margins on each car but the volume compensates for it. They've been around for 50–100 years.

The second group lacks manufacturing know-how and capacity, but they're leading the charge for technology-based changes that are required to build AVs. This group also sees themselves as the change-makers for future. Their cash-rich positions and stock-market valuations give them an edge, and some have speculated about this second group's ability to acquire OEMs, if they wanted to. But they donot want to do that because it would consume a lot of their innovative energies. They just want to leverage their computer expertise, AI skills and business models to skim the cream off the top of AV sales.

We've also seen that the first group has been acquiring smaller AI/mapping/ride-hailing start-ups to compete with the technology challengers. How the future will unfold is not very clear at all.

## 12.16 The Industry is Ready but Regulators Are Not

If you look at all the issues that need to be addressed before the auto industry can introduce fully autonomous AVs on public roads, by far the most important issue is regulations. Now, the industry has been asking for regulations for the past three years. Regulators and legislators work slowly as they have to take into account the potential reach, parameters and limitations of AV technology and consider what safeguards are appropriate, so that regulations are not over-reaching and donot slow down the advance of technology. The issue is international in scope, although one or two countries generally take the lead on such efforts and put out an initial framework. Within each country, regulations need to be established across federal, state/provincial and municipal jurisdictions. Here's my take on what needs to be addressed:

a) A broad regulatory framework at the federal and state/provincial levels has to be laid out, through a legislative tool or executive order
b) State/provincial legislation and policies have to be drafted and the intent expressed to the affected stakeholders
c) A financial framework for changes—infrastructure, administration, staff training and testing/monitoring—has to be established, including which parts will be industry's responsibility and which parts the government will agree to look after as an ongoing or ad-hoc operation
d) A process and responsibility guideline for testing, certification and ongoing monitoring of the regulatory framework has to be laid out. Would existing organizations be responsible for testing and certification of AVs? Who will fund upgrade requirements for these facilities?

In September 2017, Elaine Chao, secretary of the US government's Department of Transportation, issued a plan for AV regulations. In a document titled "Automated Driving System 2.0: A Vision for Safety," she presented Washington's new strategy for AVs. The US DOT document supports AV development and agrees that the technology will reduce accidents and human casualties. In summary, it said: do not regulate self-driving cars. The US federal position under President Trump may be at odds with the European position, which may regulate the AV industry more than the US document suggests. It's also expected that some states, like California, will have more stringent regulations than other states—similar to the way vehicle emission–related regulations have evolved. Another key piece of guidance the federal DOT gave is that most of the testing and compliance should be done voluntarily by the industry, using the best systems engineering principles and practices and observing the regulations by national and international engineering bodies such as SAE.

Here is a summarized version of the 12 key points in the US "Automated Driving Systems 2.0—A Vision for Safety" regulatory document[39]. Note that the document uses the acronym ADS (Automated Driving System) for the varying levels of autopilot capability that an AV could have. *(Source - US DOT Automated Driving Systems 2.0, 2016)*

    I. **System Safety:** Entities should follow a robust design and validation process based on a systems engineering approach with the goal of designing ADSs that are free of unreasonable safety risks. ... Entities are encouraged to adopt voluntary guidance, best practices, design principles, and standards developed by established and accredited standards-developing organizations (as applicable) such as the International Standards Organization (ISO) and SAE International. ...

    II. **Operational Design Domain:** Entities are encouraged to define and document the Operational Design Domain (ODD) for each ADS available on their vehicle(s). ... The ODD would include the following information at a minimum to define each ADS's capability limits/boundaries:

    a. Roadway types (interstate, local, etc.) on which the ADS is intended to operate safely;
    b. Geographic area (city, mountain, desert, etc.);
    c. Speed range;
    d. Environment in which the ADS will operate (weather, daytime/nighttime, etc.); and
    e. Other domain constraints.

An ADS should be able to operate safely within the ODD for which it is designed. In situations where the ADS is outside of its defined ODD or in which conditions dynamically change to fall outside of the ADS's ODD, the vehicle should transition to a minimal risk condition. For a Level 3 ADS, transitioning to a minimal risk condition could entail transitioning control to a receptive, fallback-ready user. In cases the ADS does not have indications that the user is receptive and fallback-ready, the system should continue to mitigate manageable risks, which may include slowing the vehicle down or bringing the vehicle to a safe stop. ...

III. **Object and Event Detection and Response (OEDR)**: Entities are encouraged to have a documented process for assessment, testing, and validation of their ADS's OEDR capabilities. When operating within its ODD, an ADS's OEDR functions are expected to be able to detect and respond to other vehicles (in and out of its travel path), pedestrians, bicyclists, animals, and objects that could affect safe operation of the vehicle.

An ADS's OEDR should also include the ability to address a wide variety of foreseeable encounters, including emergency vehicles, temporary work zones, and other unusual conditions (e.g., police manually directing traffic or other first responders or construction workers controlling traffic) that may impact the safe operation of an AV.

... Entities are encouraged to have documented processes for the assessment, testing, and validation of competencies a variety of behavioral competencies for their ADS ... including keeping the vehicle in a lane, obeying traffic laws, following reasonable road etiquette and responding to other vehicles or hazards. ...

**Crash Avoidance Capability:** Entities are encouraged to have documented processes for assessment, testing, and validation of crash avoidance capabilities. [This should include] pre-crash scenarios that relate to control loss, crossing-path crashes; lane change/merge; head-on and opposite-direction travel; and rear-end, road departure, and low-speed situations such as backing and parking maneuvers. ...

IV. **Fallback:** Entities are encouraged to have a documented process for transitioning to a minimal risk condition when a problem is encountered or the ADS cannot operate safely. ADSs operating on the road should be capable of detecting that the ADS has malfunctioned, is operating in a degraded state, or is operating outside of the ODD. Furthermore, ADSs should be able to notify the human driver of such events in a way that enables the driver to regain proper control of the vehicle or allows the ADS to return to a minimal risk condition independently.

Fallback strategies should take into account that, despite laws and regulations to the contrary, human drivers may be inattentive, under the influence of alcohol or other substances, drowsy, or otherwise impaired. ...

In cases of higher automation [SAE level 5] in which a human driver may not be available, the ADS must be able to fallback into a minimal risk condition without the need for driver intervention. ...

V. **Validation Methods:** Entities are encouraged to develop validation methods to mitigate the safety risks associated with their ADS approach. ... Test approaches may include a combination of simulation, test track, and on-road testing. Testing may be performed by the entities themselves, but could also be performed by an independent third party. ...

VI. **Human Machine Interface:** In a Level 3 vehicle, the driver always must be receptive to a request by the system to take back driving responsibilities. ... Considerations should be made for the human driver, operator, occupant(s), and external actors with whom the ADS may have interactions, including other vehicles (both traditional and those with ADSs), motorcyclists, bicyclists, and pedestrians. ... In vehicles where an ADS may be intended to operate without a human driver or even any human occupant, the remote dispatcher or central control authority ... should be able to know the status of the ADS at all times. ...

VII. **Vehicle Cybersecurity:** Entities are encouraged to follow a robust product development process based on a systems engineering approach to minimize risks to safety, including those due to cybersecurity threats and vulnerabilities. ... NHTSA encourages entities to document how they incorporated vehicle cybersecurity considerations into ADSs. ... Entities are encouraged to report to the Auto-ISAC [Automotive Information Sharing and Analysis Center] all discovered incidents, exploits, threats and vulnerabilities from internal testing, consumer reporting, or external security research as soon as possible. ...

VIII. **Crashworthiness:** Given that a mix of vehicles with ADSs and those without will be operating on public roadways for an extended period of time [I predict until 2050], entities still need to consider the possible scenario of another vehicle crashing into an ADS-equipped vehicle and how to best protect vehicle occupants in that situation. ...

IX. **Post-Crash ADS Behavior:** Entities engaging in testing or deployment should consider methods of returning ADSs to a safe state immediately after being involved in a crash. Depending upon the severity of the crash, actions such as shutting off the fuel pump, removing motive power, moving the vehicle to a safe position off the roadway (or safest place available), disengaging electrical power, and other actions that would assist the ADSs should be considered. ...

X. **Data Recording:** Learning from crash data is a central component to the safety potential of ADSs. ... Currently, no standard data elements exist for law enforcement, researchers, and others to use in determining why an ADS-enabled vehicle crashed. Therefore, entities engaging in testing or deployment are encouraged to establish a documented process for testing, validating, and collecting necessary data related to the occurrence of malfunctions, degradations, or failures in a way that can be used to establish the cause of any crash. Data should be collected for on-road testing and use, and entities are encouraged to adopt voluntary guidance, best practices, design principles, and standards issued by accredited standards developing organizations such as SAE International. ...

XI. **Consumer Education and Training**: Education and training is imperative for increased safety during the deployment of ADSs. Therefore, entities are encouraged to develop, document, and maintain employee, dealer, distributor, and consumer education and training programs to address the anticipated differences in the use and operation of ADSs from those of the conventional vehicles that the public owns and operates today. Such programs should consider providing target users the necessary level of understanding to utilize these technologies properly, efficiently, and in the safest manner possible.

Entities, particularly those engaging in testing or deployment, should also ensure that their own staff, including their marketing and sales forces, understand the technology and can educate and train their dealers, distributors, and consumers. ...

XII. **Federal, State and Local Laws:** Entities are encouraged to document how they intend to account for all applicable Federal, State, and local laws in the design of their vehicles and ADSs. Based on the operational design domain(s), the development of ADSs should account for all governing traffic laws when operating in automated mode for the region of operation. For testing purposes, an entity may rely on an ADS test driver or other mechanism to manage compliance with the applicable laws.

In certain safety-critical situations (such as having to cross double lines on the roadway to travel safely past a broken-down vehicle on the road) human drivers may temporarily violate certain State motor vehicle driving laws. It is expected that ADSs have the capability of handling such foreseeable events safely; entities are encouraged to have a documented process for independent assessment, testing, and validation of such plausible scenarios.

Given that laws and regulations will inevitably change over time, entities should consider developing processes to update and adapt ADSs to address new or revised legal requirements.

## My Views on "Automated Driving Systems 2.0—A Vision for Safety"

I think the DOT guidance is a reasonable preliminary effort as a policy document. The key message is that the DOT wants to encourage the industry but not get in its way with tough regulations on day one. If the industry were to follow the guidance, putting the public interest ahead of corporate interests, it would be great.

But the reality is that industry players will do what's in their own interests. Examples abound: General Motors, hauled in front of US legislators in 2014 for obvious disregard for safety in its products; VW, which violated its customers' trust by faking anti-smog numbers in government-mandated tests in the interest of its corporate performance numbers.

The DOT document sends the wrong message to the marketplace and the public by making adherence to the guidelines voluntary. There's no provision for independent third-party testing and certification facilities. No penalties are suggested for non-observance. To the uninformed public, the government will say that existing laws and safety requirements still apply, while the industry interprets the guidance in its own favor. Frankly, we need more detail, more meat and more specific rules. There should be greater insight and thought put into such a disruptive industry. I hope the European Union takes on some leadership in the regulatory environment. This is a matter of life and death.

## 12.17 AVs Disrupt the Service, Licensing and Monitoring Workforce

The arrival of AVs will have a major impact on a number of workforces in our society. The following three categories of people will be affected:

**Independent service technicians**: There are thousands of independent service technicians who service vehicles outside the dealer network of OEMs. In the future, there will be greater reliance on the dealer network because the car will become a complex electronic machine into which independents will have much less insight.

**Registration and licensing**: Hundreds of thousands of workers will have to be retrained in the regulations, business processes and operational issues that will inevitably change in the auto industry.

**Public safety officers (traffic police)**: Just as with DMV workers, traffic police will have to learn to manage the new technology on the road.

## Impact of Various Challenges on the AV Adoption Rate

In Table 12.2, I've summarized all the challenges that AVs face in terms of their impact, and also indicated whether these issues are likely to be resolved before they become critical.

|   | Challenge/Issue | Impact on Adoption | Chance of Being Resolved Before Becoming Critical |
|---|---|---|---|
| 1. | Maturity of AV technology | High—would delay widespread adoption of level 5 | Will take several iterations and revisions before technology reaches operational level or steady state |
| 2. | Disruptive nature and complexity | Very high; a show-stopper | Adoption of level 5 will likely be restricted to protected (geo-fenced) areas |

|  | Challenge/Issue | Impact on Adoption | Chance of Being Resolved Before Becoming Critical |
|---|---|---|---|
| 3. | AVs cannot see as clearly as the human eye | Medium for levels 2–4, high for level 5 | Need better resolution and new AI advances so that AVs can perceive surroundings close to how a human brain does and translate that info into actionable knowledge |
| 4. | The handoff conundrum – very important | Regulators should provide guidance | Will be resolved in next five years; minor delay in level-5 AVs |
| 5. | Ability of AV to handle severe weather conditions | Low to medium | Need more testing in cold and bad weather; Need better sensors and regulator- guidance for abandonment of AVs in bad weather |
| 6. | Robo-driver needs help under unusual road/traffic d situations | Low | AI algorithms will catch up in 6-8 years |
| 7. | Connected-car issues—the standards debate | Medium for level 5 | Will delay adoption till connectivity is standardized; will be implemented in second generation of AVs |
| 8. | No unified HMI (Human Machine Interface) | High | Will delay adoption significantly. Industry will ultimately adopt common HMI |
| 9. | Cybersecurity issues | Very high | Will delay adoption—expect some hacking during first phase |
| 10. | Not prepared for hybrid environment | High | Will be resolved in phase 2 and 3 but will delay adoption |
| 11. | City infrastructure not ready | High | New AV taxes may have to be introduced to pay for this; will take a long time |
| 12. | Business models speculative and optimistic | High | Sensor prices and cost structure of AV will have to come down |
| 13. | Marketing challenges | High | Business knows how to handle this one |
| 14. | Nobody holds all the marbles | Medium | Business as usual—no change |
| 15. | Industry ready—regulators not | Medium | Ultimately the public will force legislators will act |
| 16. | Training new breed of AV-savvy workforce | Medium | The industry and government departments will adapt and train – will take time |

*Table 12.2 – Major Challenges Facing AV Plus Industry*

**Summary**

*I have described in this chapter the challenges that the AV industry faces. In a free economic system, there's no single body in charge of addressing these issues. The self-regulating* mechanism works slowly. Industry innovators must address those challenges that they can. Regulators and legislators must pitch in. In due course, it will happen. The question is, when?

## Citations for External Content References

[31] New York Times Article- https://www.nhtsa.gov/sites/nhtsa.dot.gov/files/documents/13069a-ads2.0_090617_v9a_tag.pdf

[32] US Government Accountability Office GAO-16-350) Cybersecurity Issues https://www.gao.gov/assets/680/676064.pdf

[33] Todd Litman Report - https://www.vtpi.org/avip.pdf

[34] Stanford CARS tackles handover problem – https://news.stanford.edu/2016/12/06/taking-back-control-autonomous-car-affects-human-steering-behavior/

[35] Volvo's view on AV fallback problem - https://www.wired.com/2017/01/HUMAN-PROBLEM-BLOCKING-PATH-SELF-DRIVING-CARS/

[36] Severe weather conditions - https://www.freep.com/story/money/cars/mark-phelan/2018/04/01/detroit-self-driving-car-center/472061002/

[37] Drive.ai - https://www.drive.ai/

[38] Report on by US Government Accountability Office on AV cybersecurity- https://www.gao.gov/assets/680/676064.pdf

[39] 12 key points in the US "Automated Driving Systems 2.0—A Vision for Safety" regulatory document https://www.nhtsa.gov/sites/nhtsa.dot.gov/files/documents/13069a-ads2.0_090617_v9a_tag.pdf

# Chapter 13

# Opportunities for Entrepreneurs and Startups

*A book on AV would not be complete if it did not provide encouragement to entrepreneurs and start-ups to find a niche area where they might contribute. Having founded the Toronto chapter of TiE—the largest network of entrepreneurs in the world—I feel that I should outline some areas of the AV race that could encourage more entrepreneurs help the industry do it right.*

The entire landscape of the auto industry is going to change over the next few decades. The AV industry is only in the first phase of its development. I said earlier that the AV industry could have a lifespan of 100 years before another major disruption takes place, and we are only in its first decade. There are still many unsolved problems in the AV sector. Innovators have opportunities to create solutions not only in the core area of the AV itself but also in the ancillary and support areas. In this chapter, I'll highlight the opportunities for researchers, innovators and start-ups to address some of the unsolved or partially-solved problems in the following areas:

1. Research for AVs Plus
2. Cheaper and more functional sensors
3. Software enhancements, especially in the AI autopilot
4. HMI and OS
5. AV Plus connectivity services
6. Cybersecurity services
7. Smartphone integration, infotainment content and AV applications
8. Mapping and localization services
9. Support apps, e.g., scheduling apps, parking-space apps
10. Fleet operations and management apps
11. Next-generation transportation infrastructure—traffic lights, road signs, etc.

## 13.1 Research Areas

Several leading universities in the United States, Germany and other countries have been doing leading-edge research into AVs since 1975. Notable among these are CMU in Pittsburgh, MIT in Massachusetts, University of Stanford in California, the University of Michigan, Bundeswehr University in Munich, the University of Parma in Italy and Tsinghua University in Beijing. More recently, many other universities around the world have created a chair for AV-related research. The following areas still need academic attention and research focus—some at the basic level and some at the application level, i.e., commercializing the basic research that's already been done:

- Better computer-vision processing using the current and future set of sensors
- Radar and other sensors with wider FOV (field of view) and longer distance

- Less expensive "low light" computer-vision cameras for night and poor-visibility periods
- More efficient (from a processing point of view) sensor fusion—the ability to discard redundant info from multiple sensors and consolidate only the relevant information for a composite vision of the terrain at the application level
- More advances in AI—going beyond Geoffrey Hinton's neural and capsule networks. AI is still unable to accurately mimic what our brain perceives from the images our eyes see. Sensors and cameras collect a lot of information but our AI-based processing of all the sensor data still doesn't give as accurate an interpretation of the surroundings as our brain does from the relatively smaller amount of information collected by our eyes. Science still may not know accurately how our brain does this in good weather, in bad weather and in low light. The gap is in deep learning, the processing of tremendous amounts of data, the creation of a virtual perception of the surroundings and the use of it to make decisions that guide the movement of the AV—no small feat! (As I said before, I'm not an AI expert, but that's how I understand the current state of the art. Please see my comments in Chapter 12).

CMU has also put out a report that is quite instructive, entitled "[Autonomous Car Policy Report](#)."
Among other things, it points out another major decision-making challenge for the robo-autopilot: not knowing the intentions of other vehicles (whether they're AVs or non-AVs operated by humans).

There are simply a lot of opportunities to improve AI for AVs. There are lots of holes in the current versions of autopilot, whether from Waymo, Tesla or the Silicon Valley R&D labs of the established OEMs. Ask the experts at CMU and CARS and they'll tell you that there's a long way to go before the fully autonomous SAE level-5 AV meets acceptable Sigma 5 quality-control standards for human interventions. Of course, if the DOT and the NHTSA accept that we shall sometimes allow scores of AVs without human drivers to fail on busy national highway convergence points with multiple stranded occupants in each AV, that's a different matter. I am not sure if Waymo or anybody else has simulated such emergencies in their labs or established procedures for getting everything back in order and moving in a short time.

Figure 13.1: *Autopilot makes its way into mainstream comics (Source: IToons and The Times of India)*

## 13.2 Cheaper and More Functional Sensors

Sensors are the underpinnings of AVs. Each AV requires multiple sensors for three reasons—we need 360-degree coverage, the sensors have limited range and no single sensor gets all the info autopilot needs. The industry is still trying to figure out what configuration of sensors is optimal from functional and cost points of view. There are opportunities to combine the functionality of multiple sensors into one to reduce complexity and cost. This can be achieved through engineering research, development and testing. I believe sensors will evolve over the next five years. Start-ups can go through patent-office submissions to uncover opportunities to develop better sensors.

Two big areas in sensors are LiDAR and low-light cameras for AV applications. The cost of sensors, especially LiDAR sensors, is still very high and must come down. Google and Velodyne have proposed a solid-state LiDAR that will cost less than current models. Hardware start-ups in this area can stimulate innovation and competition resulting in lower component costs and ultimately leading to a lower cost differential between AVs and non-AVs.

As many as 50 start-ups have tried to raise money for LiDAR sensors in 2016–2017 with VCs actually funding $600–700 million, according to press reports. Tier-one component suppliers like Continental, Bosch, Delphi and Magna have gotten into the act as well. The LiDAR sector is almost saturated, unless somebody comes up with a unique design concept.

## 13.3 Software Enhancements, Especially in AI portion of the Autopilot

There are still a number of driving situations that have not been addressed by the current robo-autopilots on trial by established OEMs and technology challengers. While innovative work is going on in these areas, here are some examples where more innovation is required:

- The handover conundrum (discussed in Chapter 12), which will be important until we've built an autopilot that's fully capable of handling any and every situation reliably
- Parking in very tight spots like a cluttered home garage—should a level-5 AV be parked by a human driver and, if so, how will that work if there are no pedals?
- Communication with other drivers (human or autopilot) in the vicinity, as well as with pedestrians
- Should autopilot *honk*, and in what situations?
- The whole protocol of communication with a remote support center to make sure AVs in trouble—those who have given up and won't move—don't shut down a road
- How should autopilot deal with driving in the rain, slush, ice and snow—should it be through environmental customization using a "settings"-type approach or should we expect the autopilot to know how to adjust for the weather like a human driver does?
- Ethical issues in driving (see Chapter 11)
- Self-customization for different styles of driving—passive and safe, active and aggressive

**Drive.Ai**—a Silicon Valley start-up—is trying to address how an AV will communicate with pedestrians to indicate when it is safe to walk. Similarly, an AV will need to have a system for understanding and replicating the gestures and eye-contact that humans often used to communicate with one another, such as a driver requesting access to change lanes or when a traffic cop is directing traffic despite functioning street lights. Some of these situations have been addressed but there are others that have not yet been tackled. How these questions get integrated into different autopilot software solutions is something that the industry must address. These are sort of like sub-routines in scientific programing, which should be capable of integration into a general-purpose autopilot.

## 13.4 HMI and OS

The AV Plus is becoming a complex multi-computer environment that has grown out of established OEMs developing standalone solutions for specific tasks using task-centric chips and computers. This was fine for non-AVs where we could control task-oriented computers through switches and toggles, but in AVs, everything is software-controlled at a higher level. So this presents a user-interface issue that has become quite complex.

My view is that there are companies in the AV landscape that have different design expertise and philosophies. The OEM computer experts have one philosophy and the technology challengers have another. What we need is someone to bring these philosophies together to create an integrated, unified approach. OEMs don't have much OS development expertise, yet they want to develop their own unique operating environments. In contrast, technology challengers like Waymo have deep expertise in developing computer operating systems but may not fully appreciate the variety of task-oriented chips and controllers that need to be tied together in an AV. They may want to control everything through a modern interface that might not satisfy classic car drivers who want a combination of toggle switches and icons. A hybrid multi-mode HMI that can be customized through a smartphone-like "settings" concept may work better to satisfy multiple groups of potential AV adopters. HMI innovators may be able to come up with unique solutions to this need.

## 13.5 AV Plus Connectivity Services

An important attribute of an AV Plus will be its connectivity. This is an area that is wide open—no single vendor has emerged as a leader. The industry will have to create hardware modules for short-range V2V communication and long-range V2I communication. Qualcomm and NXP are very active in this area. Using these hardware modules, innovators can create "plug and integrate" software modules that they can sell to OEMs and other AV manufacturers. Innovators can become familiar with DSRC and 5G standards in the AV context to create niches for their connectivity offerings.

## 13.6 Cybersecurity Services

Whenever you connect a vehicle with other vehicles in the vicinity through (V2V communication) based on DSRC standard or use public 5G networks for V2I communication to update maps (downloading the latest changes to the roads as well as crowd-updating information [with tools like Waze] on current events like construction work), you open up the possibility that hackers might enter the network. This is a serious concern that will definitely delay AV adoption if it's not addressed to consumers' satisfaction. The field is wide open for innovators to come up with solutions that are compatible with the chosen operating systems. Where the OS selected by an OEM is proprietary—as in most situations—the solutions must be open-ended so they can be integrated.

Many entrepreneurs and investors are becoming active in AV cybersecurity. As reported by Techcrunch[40], an Israeli company called Argus Cyber Security[41] raised $26 million and auto electronics maker Harman paid $72.5 million for cybersecurity start-up TowerSec in 2016. More AV cybersecurity start-ups are appearing every day.

## 13.7 Smartphone Integration, Infotainment Content and AV Applications

One of the key attributes of a future AV Plus is that it will be capable of providing a platform for full access to entertainment streaming services and be integrated with Apple CarPlay, Android Auto and similar applications. It will have Wi-Fi hub support so that occupants can access the internet through 4G today and 5G tomorrow.

There are lots of creative applications that app writers can develop for the AV. In fact, most of the smartphone applications we have today should have an AV version in the future. Since the AV screen is bigger (though with lower resolution) than the smartphone screen, it'll make sense to rewrite various applications for larger screens. Right now, there are no device standards or network standards, so the whole area of AV app underpinnings is quite hazy... but innovators can start working on it.

## 13.8 Mapping and Localization Services

One of the key pieces of software that an AV uses is its mapping and localization software. In order to plan a route and navigate, the car must have a detailed, up-to-date map of the area. In fact, now what AVs really need are high-definition 3D maps including all the road signs, lane markings, painted signs, road-edge obstructions and other characteristics that might be needed to facilitate autonomous driving. Creating high-definition digital maps and keeping these maps up-to-date is a difficult task.

There are already several leading companies doing this, such as HERE (a German company, which was acquired by a consortium of Audi, BMW and Daimler in 2016) and American company TomTom. Once the base specifications are standardized (de facto across the industry in the next few years) or designed in proprietary systems (as they are now), we'll need an army of mapping crews to go out and map the world. Google did this for Google Street View, but the requirements have changed for AVs and we'll need to do this again. It will happen slowly and gradually, and there's lots of room for innovators to get involved.

## 13.9 AV Support Apps—Parking Space Locations, On-Demand Pick-up Scheduling

There are a number of support applications for tomorrow's AV world. We'll need an inventory of parking spaces available in real time. Every parking lot that wants to be considered for in-out rental on short- or long-term bases is going to have to update its inventory through new software, and someone will have to figure out how to centralize this information in a server that is available to TaaS providers as well as private AV owners.

On-demand scheduling of pick-up clients requires sophisticated software that can create optimal routes. This type of software does exist, such as ride-scheduling services for people with disabilities in various cities. This could be useful for municipalities where small fleets of AVs are used to pick up passengers from their homes or designated pick-up spots. But the entire landscape will change and require automated booking and scheduling applications.

## 13.10 Fleet Operation and Management

AVs are expected to discourage private ownership of cars because on-demand TaaS will become so cheap and efficient, provided consumers are willing to share rides with others. As the desire for individual car ownership declines with every new generation, TaaS fleets will become more and more common. As a result, we will need new fleet operation and management applications developed for the AV world. While there are several such applications already, an entirely new generation of these apps will be required that can interface with AVs.

A new category of fleet operators along the lines of Tesloop, RideCell and Navya (see Chapter 8) will emerge. One recent example called Voyage services senior residents of retirement homes in Florida and in San Jose, California, using a dedicated fleet.

## 13.11 Next-Generation Traffic Lights and Road Signs

As we focus on upgrading our road infrastructure, there will be an opportunity to create standards and build chips for smarter traffic lights and road signs that can communicate with AVs.

## 13.12 The AV Landscape[42]

As I've said several times in this book, the auto industry is large, complex and multi-layered. It is becoming even more complex with the switch to autonomous vehicles. From the investors' and innovators' viewpoints, it can be partitioned into many segments. Figure 13.1 (from VentureBeat) shows one way to segment the AV landscape.

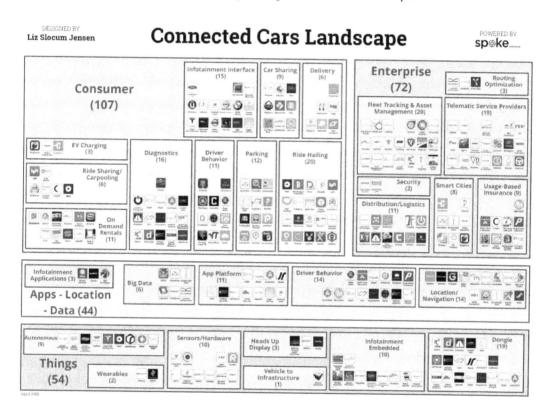

*Figure 13.1: One way to segment the extremely complex AV landscape*
*(Source: Liz Slocum Jensen of VentureBeat; used with permission)*

This figure designed by Liz Jensen of VentureBeat in Silicon Valley, shows the incredible width and breadth of the AV landscape. The resolution of this graphic makes it impossible to appreciate the details, but the point it makes is crystal clear.

## Summary

*I have mentioned many areas, either core parts of AVs or on the periphery of AV Pluses, where there are opportunities for innovators who think outside the box and aren't scared off by initial failures or lack of current knowledge. Humankind has progressed because we discovered what was unknown and solved problems that could not be solved, and we'll do so here again.*

## Citations for External Content References

[40] Techcrunch on AV cybersecurity - https://www.crunchbase.com/organization/towersec

[41] https://news.harman.com/releases/harman-to-acquire-towersec-automotive-cyber-security

[42] Source – Virtual Beats – Liz Slocum's chart

# Chapter 14

# Crystal-Ball Gazing—What, When, How and the Evolutionary Path

> *Over the past two years, a lot of excitement has been created and expectations built up in the minds of the public about autonomous vehicles. With extensive media coverage of AV trials taking place in the developed world, the public is asking: what are we getting, when is this becoming real and how will we get there.*

I have covered a fair bit of ground in the book so far. First I discussed AV fundamentals, AV history, AV rationale, AV technology framework and the TaaS business concept. Then I looked at AV landscape, stakeholder perspectives and challenges the industry faces. Now in the last and final part of the book, I want to indulge in some crystal-ball gazing, and try to make some predictions about how long we'll have to wait till most of us are riding in AVs. Since we won't get there in one single step, I'll also speculate on the most likely path the AV might follow en route to widespread adoption.

## 14.1 What Incremental Value Is the Consumer Getting in an AV, and at What Price?

Consumers are asking, "What is it that an AV will give me personally compared to a luxury car with ADAS features?" Societal benefits—the possible reduction in traffic congestion and the environmental advantages—are important to the public in general, although somewhat secondary to the individual consumer. They *will* look at price differences. The price point will be crucial in determining the rate of adoption of AVs.

It's common practice for marketing executives to set the initial price of a new, disruptive innovation quite high for several reasons:

1. This high price gives customers sense that the additional features are highly valuable.
2. It sets a ceiling for the price, so that when the price point eventually comes down, it gives customers the illusion that they're getting a deal.
3. Initially, suppliers want to attract customers who are not price-conscious.
4. The cost of components going into an upgraded product actually *is* high in the early stages.
5. Manufacturers want to recover their R&D costs as soon as they can.

How do we measure, and how will *consumers* measure, the delta improvement in features and benefits from a conventional car to an AV? We looked at this topic in Chapter 4 from society's point of view but let's look at it from a consumer's self-interested perspective.

The basic functionality—getting us from point A to point B—already exists in the conventional vehicle that the consumer already owns. The major benefits include the excitement of self-driving, the economics, reduced stress, improved safety and extra time for productivity or pleasure.

What this differential in price between conventional non-AVs and AVs would be on day one when these cars are introduced is any body's guess because the auto OEMs are not giving any indication. Intelligent drive feature on my Mercedes C300 cost around $3K but it does not have the active-assist features that you really want. I am speculating that $15K is a reasonable differential to expect, when dust settles on the ground.

We have to guess at what the differential in price will be on the day AVs are introduced, because OEMs are not giving any indication. The Intelligent Drive feature on my Mercedes C300 cost around $3,000 but it does not have the active-assist features that you want. I am speculating that $15K is a reasonable differential to expect. I speculate that $15,000 is a reasonable differential to expect for a level 5 AV.

## 14.2 When Can We Expect AVs?

"When" is one of the most-covered topics in the media.

### 14.2.1 Predictions by the Auto Industry

Let's look at some of the recent headlines announcing date of introduction of AVs (with thanks to www.driverless-future.com for assembling much of this list).[43]

**Elon Musk clarifies Tesla's plan for level 5 fully autonomous driving: 2 years away from sleeping in the car**, Source *Electrotek.co*, *April 29, 2017*

Tesla's CEO clarified his plans in April 2017 and predicted that true level-5 autonomy was about two years away. This suggests 2019 is Tesla's target.

**CES 2017: Nvidia and Audi Say They'll Field a Level 4 Autonomous Car in Three Years**, *IEEE Spectrum*, *January 5, 2017*[44] — Scott Keogh, head of Audi America, announced at CES 2017 that an Audi that really would drive itself would be available by 2020.

**Volkswagen Readies its Robo-Fleet**, *Autonews.com*, *September 17, 2017*[45]

Johann Jungwirth, Volkswagen's head of digitalization strategy, expects the first self-driving cars to appear on the market by 2021.

**Ford CEO Announces Fully Autonomous Vehicles by 2021**, *Reuters*, *Apilt 29, 2016*[46]
Mark Fields, Ford's CEO at the time, announced that the company plans to offer fully self-driving vehicles by 2021. Initially, these vehicles, which will have no steering wheel or pedals, will be for TaaS. Fields said he expects that it will take several more years before Ford will sell AVs to the public.

**GM to Launch Autonomous Cars in Big Cities Sometime in 2019,** *CNBC*, November 30, 2017[47]

General Motors expects to have autonomous vehicles working commercially in big cities sometime in 2019.

**Andrew Ng: When will self-driving cars be on roads?** *Quora*, January 29, 2016[48]

In an interview session, Andrew Ng, the chief scientist of the Chinese search engine Baidu, said he expects that a large number of self-driving cars will be on the road within three years, and that mass production will be in full swing by 2021.

**Toyota to launch first driverless car in 2020,** *Wired.com*, October 8, 2015

Toyota is starting to overcome its long-standing reluctance with respect to autonomous driving: It plans to bring the first models capable of autonomous highway driving to the market by 2020.

**Driverless taxi firm eyes operations in 10 cities by 2020,** *Yahoo News*, August 29, 2016[49]

NuTonomy started trials of its self-driving taxis in Singapore in 2017. It plans to expand to 10 cities by 2020.

**Autonomous cars will arrive within 10 years, Intel CTO says**, *Computerworld*, October 22, 2012[50]

Justin Rattner, CTO of Intel, predicts that driverless cars will be available within 10 years. Intel is hoping to equip autonomous smart cars with its Atom and Core processors.

**Look Ma, No Hands!** *IEEE*, September 5, 2012[51]

IEEE estimates that up to 75% of all vehicles will be autonomous by 2040.

## 14.2.2 Predictions by Researchers (CMU, Princeton and Berkeley)

Serious researchers are not as optimistic as the industry representatives in their predictions. Here's what a few leading research scientists have said.

"It's that last 10%, and then the last 1% of circumstances that gets very, very difficult to solve," Andrew Moore, dean of CMU's School of Computer Science, said in a recent interview. "Autonomous in-town driving, I'm still putting it at 2028" as published in Forbes Sep 28, 2016[52]

Steven Shladover of the University of California, Berkeley, who has worked on automated driving for more than 20 years, says, "Probably what Ford would do to meet their 2021 milestone is have something that provides low-speed taxi service limited to certain roads—and don't expect it to come in the rain.... The hype has gotten totally out of sync with reality." Source – MIT Technology Review [53]

Alain Kornhauser, a Princeton professor and director of the university's transportation program, also expects 2021's vehicles to be very restricted. "By then we may be able to define [a] 'fenced' region of space where we can in fact let cars out there without a driver," he says. "The challenge will be making that fenced-in area large enough so that it provides a valuable service. Source – "MIT Technology Review [54]

### 14.2.3 Opinions by Major Consulting Firms

The analysis by the major consulting companies recognizes both market and non-market factors and then provides general guidance to the marketplace. Here's my interpretation of what these firms are saying.

**McKinsey & Company** conducted independent research, both on its own and in combination with Stanford University. McKinsey's 2016 report ("Disruptive trends that will transform the auto industry") suggests two scenarios—high-disruption change and low-disruption change. Under the high-disruption scenario, AVs could arrive in 2025 in a modest way (2% of new sales) and reach 15% of new sales by 2030. The high-disruption scenario entails regulator challenges overcome in key markets (the United States, Canada and Europe), safe and technical solutions fully developed and consumers enthusiastic and willing to pay the premium. Under the low-disruption scenario, McKinsey predicts only 2% uptake for level-4 AVs by 2030 and 10% uptake for level-4 AVs by 2040.

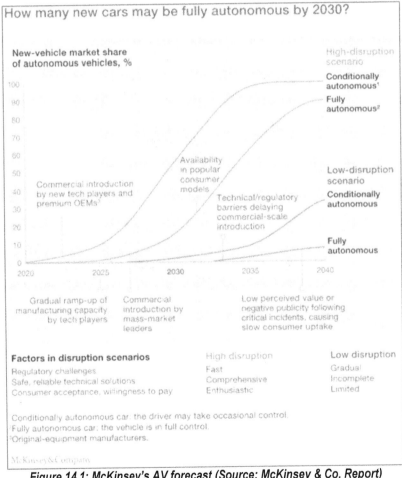

*Figure 14.1: McKinsey's AV forecast (Source: McKinsey & Co. Report)*

When I merge the findings from the McKinsey report with market forecasts from the research community and other consulting companies, I tend to lean more toward the low-disruption scenario as the most likely outcome.

**Deloitte** has conducted extensive consumer-preference surveys and produced several research reports ("2016 What is Ahead for Autonomous Driving"55 and "2018 Global Automotive Consumer Study"56) that support some of my conclusions regarding the lack of enthusiasm for fully autonomous cars. I discussed Deloitte's survey findings in Chapter 9. I believe that because of the sample size (20,000 consumer responses), Deloitte's findings are likely highly accurate as of 2016, when the survey was conducted. 2018 study reflected greater interest in AVs. I recognize that the fluidity of the AV market and fast-paced technology developments could change the results with time.

**AT Kearney**, in its AV market report ("How Automakers Can Survive the Self-Driving Era"), estimated that the global market for automated and autonomous driving (including ADAS, AVs and associated services) could be 7% of the total automotive market in 2030 and 17% in 2035. These findings are more conservative than the forecasts of AV suppliers in general but more optimistic than McKinsey's.

**KPMG**'s report ("Self-driving cars: The next revolution") makes predictions that are also generally in line with my overall conclusions that it will take much longer than what the auto companies are leading us to believe for AV adoption on a significant scale.

To be fair to the auto suppliers, they have not indicated how many AVs they will deliver to ride-hailing or TaaS service providers by the announced dates. They could supply AVs fully capable of self-driving (according to their own benchmarks) on some (but not all) public roads, but TaaS providers like Lyft, GM and Uber could decide to operate these vehicles with a human driver for the first few years as they iron out all the kinks and gain customer trust.

### 14.2.4 Opinion of Road Infrastructure Policy Advisors

Todd Litman, executive director of the Victoria Transport Policy Institute, published a detailed report in 2017 in which he comments on the major benefits cited by AV promoters (i.e., reduction in traffic congestion, accidents and casualties). He believes that many of these benefits are over-stated and too optimistic. Litman concludes that actual experience during 2021-2030 will be less dramatic than what is forecast. AVs may operate at the designated speed on divided highways in good weather but will operate at lower speed than regular traffic in congested areas. In the first decade, cost of TaaS (private) will probably be $0.80 to1.00 per mile and 30-60 cents per mile for shared TaaS – he calls it micro-transit. Shared TaaS will provide lower quality ride and customer service than human-driver operated taxi or Uber/Lyft services. AVs will continue to be unreliable in heavy rain and snow storms. He also says that it will be at least 2040 when AVs will be affordable for middle income consumers. Litman concludes "despite the low price of self-driving services, many travellers will pay extra for higher-quality human-operated taxis and public transit." **I generally agree with Mr. Litman's overall assessment.**

## 14.3 AV Stakeholder Forecasts: All Over the Map

We are in a period of AV development where OEMs and technology challengers are making claims and predictions without explaining their underlying assumptions. This is not uncommon at this stage, because consumers don't know any better and aren't questioning the proponents. But it's important to look at the details so we can arrive at some kind of clarity.

OEM forecasts do not specify the types of highways/roads or geographical areas where their early AVs are going to operate. The DOT document I quoted in Chapter 12 suggests that such operational design domain (ODD) parameters should be spelled out. Even if spelled out, will these appear in the advertised specifications and sales documents, or will they be buried in fine print in some obscure place? It seems to me that an AV could satisfactorily meet SAE level-5 objectives in a protected, geo-fenced area of a specified city while not meeting the same designation on a public street in Manhattan.

Let us illustrate the ambiguity with an example. Let's say that sometime in 2019, a level-4 AV is able to travel on the US I-90 from Boston to Seattle, self-driving 95% of the time with 15 human interventions over a 3,000-mile route. Does this car qualify for SAE certification? SAE level-4 specifications give no criteria for how many times a human driver can be asked to intervene over a given period and for how long.

The readers may have noted that Elon Musk has promised that his Tesla model S will self-drive from Los Angeles to New York in early 2018. Elon has not provided rules that his car will follow—whether there will be a human driver in the car or in another car following the autonomous car undergoing the test. The problem I see is that there is no referee in this game, and Musk is setting the rules as he sees fit. It's a free-enterprise AV rally.

**Setting Quality Control Standards for the Autopilot**
It is pretty difficult to prescribe quality control standards for the AV industry in so far as the autopilot is concerned. First of all, you need to decide one or more metrics that can be measured and define quality of autopilot performance. In my view, there are two obvious candidates for this metric – one is the number of alerts for human intervention and second is the number of deaths during its operation over a defined period or miles travelled. Airline industry has set the number of deaths for passenger miles as a metric for measuring its performance. It has decided that the industry must not exceed one death per two billion passenger miles for meeting 6-sigma quality standard. Some AV academicians have suggested a similar metric for the AV industry. If the industry were to use number of human interventions, it might be 2/3 interventions for million miles of travel, based on a suggestion by a CMU professor. It might be pointed out that, according to Allstate, average driver has one accident for 165,000 miles. Another statistic that National Safety Council of the United States uses is 1.25 deaths per 100 million miles travelled.

Obviously, experts will have to discuss, debate and decide what the target number for AV should be. Needless to say that, a target quality control standard is required.

### Clarifying the Notion of "AVs Have Arrived"

Having a few hundred or even a few thousand AVs driving on geo-fenced roads with lots of restrictions would not mean that "AVs have arrived." Even a few thousand AVs on roads would still signify AVs in trial mode. As a management scientist, I want to define the notion in a more quantitative and meaningful way. I'll distinguish between SAE levels 3 and 4 from SAE level-5 AVs because there are distinct differences in terms of responsibility and control between the two categories.

### 14.3.1 My Definition of "Arrival" of Autonomous SAE Levels 3 or 4

Here's my definition of "arrival" of Level 3 or 4 AVs:

- In a given year, several (say, more than three) auto suppliers have delivered at least 100,000 SAE level-3 or level-4 AVs in North America that can make end-to-end trips (from a public garage or a private house to any destination of choice by a passenger) without a human driver in control.
- Each of these vehicles has driven 50,000 miles on average in a year. This assumes primary use is TaaS. Private use of AVs will come later.
- There are no more than 20 human-driver interventions per million miles traveled (defined as meeting the five-sigma quality-control standard, which I suggest is sufficient for the first five years, despite the six-sigma criterion suggested by CMU in an unaudited report. Six sigma quality standard translates to 2-3 interventions per million miles.
- The AVs have been certified by a third-party testing lab that is not affiliated with any vendor from a board or investment point of view.
- There is no unreasonable restriction on the use of the AVs in terms of weather where they are licensed—they are just as comfortable in variations of weather in sunny California as in snowy Alberta. We cannot come to a stand-still because of weather.
- Fatal accidental record of the AV industry is at least 5 times better than that of non-AV. This criterion translates to 0.25 deaths per 100 million miles of AV travelling.

I am **guesstimating** that the industry will reach this level of AV arrival between 2025 and 2030. I stress the term "guesstimate," which suggests the difficulty of assigning any level of probability to my estimate. However, it *is* based on my 40 years' experience as a consultant and my in-depth review of the opinions of independent experts, consulting companies and researchers. The probability that I would assign to my guesstimate is 25% in 2025 and 75% in 2030.

### 14.3.2 My Definition of "Arrival" of Fully Autonomous Level 5

The additional requirements for meeting my definition of "arrival" of fully autonomous SAE level 5 AVs are as follows:

- At least 100,000 level-5 AVs have been sold or are in use by TaaS providers in North America. Suppliers have begun offering AVs for private purchase as well.
- There are no human drivers riding TaaS AVs as backup.
- We've reached the quality control standard would be what has been proposed by academic institutions—2-3 interventions per million miles traveled.
- There is remote support to take care of AV jams on the road. AVs in trouble are moved aside or towed away through acceptable procedures.
- Like the airline industry, all critical computer components (sensors and AI servers) have "hot" standby, i.e., redundant design with a fail-safe feature—in case of failure, the OS automatically switches over to the hot standby.

I estimate that there is 50% likelihood (for the legal sticklers, I make no guarantees) that the industry will reach this level of autonomy by 2030, assuming the average price differential for fully autonomous cars doesn't exceed 15% of the price of an equivalently configured non-AV. Send me an email if I'm right (my address is at the beginning of the book).

### 14.3.3 Projections of AV Adoption to Mass Acceptance Level

Whenever vendors introduce an emerging technology product, there's an initial euphoria about that product and early adopters buy it out of newness, but that's quite different from mass acceptance. How AV adoption moves forward depends on many factors, including the following:

- How good is the initial experience of the early adopters?
- Is the price differential between AVs and non-AV in the same class equal or less than the consumer perception of benefits?
- What's the overall interest in the perceived benefits of AVs in congested cities? (A small number of AVs isn't going to make a dent in the congestion or result in fewer accidents. In fact, there may be more confusion in the beginning when we're in a hybrid environment.)
- How good are the marketing campaigns?
- Is the TCO (total cost of ownership) truly in favor of AVs?
- Are the TaaS costs significantly lower than the cost of operating a privately owned car? (I've said earlier that the TaaS business model is flawed for exclusive use of the AV—shared use is a different matter. According to my calculations, TaaS will be successful in taking over taxi services but it will not reduce private ownership of cars significantly initially till Y-Z generations become dominant adopters.)

I expect that there will be a very slow, gradual uptake of AVs in the first decade (2021–2030). In the next decade (2031–2040), uptake will pick up (to about a 10–15%

compound annual growth rate [CAGR]) because Generations Y and Z will become a significant portion of the AV customer base. In terms of numbers, I support McKinsey's "low-disruption" predictions, i.e., that AVs will make up 2% of new-car sales (300,000) in 2030, going to 10% (two million) in 2040. These numbers reflect North American (including Canada) sales—around 20 million cars on annual basis.

## 14.4 How AVs Might Be Introduced

There are two important tactics that will influence *how* AVs will be introduced. First, OEMs are stressing their incremental and gradual approach to introducing this new technology. Technology challengers are going for the ultimate target (level-5 AVs) directly. Established OEMs have started including ADAS features in their luxury vehicles and will start putting them into semi-luxury vehicles as well; then, they'll introduce level-3 and level-4 AVs; and finally they'll offer level 5s. I have to stress that introducing a level 5 for the purpose of demonstrating technology prowess is different from offering it as a strategic product with a full marketing thrust.

This approach, however, is consistent with the consumer appetite. I think, by and large, the majority of consumers will respond to the OEM strategy. Only a small minority of affluent techno-savvy consumers will buy into the strategy espoused by technology challengers like Waymo.

I believe the auto industry will introduce an AV product that is capable of SAE level-4 or level-5 functionality through TaaS providers using the ride-hailing business model. Initially, these TaaS AVs will be staffed by human drivers, who will morph into robo-drivers in due course, as suppliers and customers get comfortable with test results and the number of human interventions drops to five-sigma levels. My guess is that this will happen two to three years after the launch of the TaaS offering. This TaaS experiment will allow the industry to road-test the robo-autopilot with paying passengers and to test many variations of business and pricing models.

OEMs will also sell AVs to certain organizations such as universities, airports, hospitals, convention centers, amusement parks and airports that will use AVs to run mini-TaaS services within their own campuses/premises. They might also supply AVs to municipalities that want to replace their inefficient bus services with TaaS services at a lesser cost. The public will be happier because the wait and trip times will go down.

The second tactic that AV suppliers have suggested is that they won't sell AVs to private buyers initially. This is a wise decision based on the pragmatic reality that the product won't be consumer-ready on day one (see the first and second phases in Table 14.1). Backlash from negative feedback from private consumer sale can be disastrous in any industry especially if they have to accept liability for accidents. Therefore, AV suppliers will be extremely cautious and would not jeopardize their private car sales business.

## 14.5 The Evolutionary Path of the AV Plus Revolution

The destination of the future car is AV Plus—a fully autonomous (SAE level 5), well-connected, cleaner (electric) and smarter vehicle with an integrated user interface in which passengers can enjoy infotainment apps including smartphone integration. While the emphasis in the industry's single-minded quest is on the self-driving aspect, other attributes will be equally important in the long run.

AV Plus is a lofty target and it will take a long time for the industry to develop products that meet the AV Plus designation. It won't really, fully exist until the majority of the driving population, at all income levels, has been driving autonomous cars for as many as 40 to 50 years, in my estimation. It could be around 2060 that the majority of cars on the road in the developed world are AV Plus with no humans driving at all, if we get to that extreme. Undoubtedly, on the way there will be many twists and turns.

Table 14.1 outlines five time periods I see unfolding in the progression of the AV. For those who find this timeline too conservative, remember that my criteria are slightly different from those of most of today's AV proponents.

|   |   | Phase 1 (2017–2021) | Phase 2 (2021–2025) | Phase 3 (2025–2030) | Phase 4 (2030–2040) | Phase 5 (2040–2050) |
|---|---|---|---|---|---|---|
| 1 | Development (HW, SW, AI, OS, HMI) | Major AV development | Autopilot enhancement; OS/HMI development | Cheaper sensors; AI for level 3/level 4 certified | Autopilot for level 5 certified by 3rd parties; HMI and OS standards | HW and SW stable |
| 2 | AV Market Development | Optimistic delivery dates; acquisitions, mergers and partnerships | Partnerships undergo change; better business models | Better idea of where the AV market is going | AV adoption by Gens Y+ Z; reality sets in; vendors re-prioritize | Market stabilizes with realistic expectations |
| 3. | Market Deliverables | Level 2 exists; validation of AV level 3 to level 4; platoon trials | Level 5 consumer trust-building; platoon trials on highways | Expect level-5 delivery for private sale by 2030 | TaaS AVs sell in quantity; AVs to private owners also | 20–30% of new cars AVs (level 3–level 5) |
| 4. | TaaS | TaaS trials on geo-fenced campus routes | TaaS on selected public roads with human drivers | TaaS trials without human drivers | TaaS wipes Taxi industry, reduces private ownership | TaaS adoption increases significantly |
| 5. | Private AV Ownership | None; more Drive Me AV trials | Selective level 4s offered to fleets for trials | Limited private ownership introduced | TaaS/private ownership strategy clarifies | TaaS and private ownership co-exist |
| 6. | Strategy: OEMs1 (Leaders) | Develop Level 3 & 4, R&D for level 5; acquire & invest | Level-5 trials with TaaS (with human driver as backup) | Offer level 5 to TaaS (no human driver) | Level 5 to private owners AVs for middle consumer tier | AV Plus: smarter, greener and connected |
|   | Strategy: OEMs2 (Followers) | Respond with ADAS as first step | Level-4 and level-5 trials for TaaS; own OS and AI or license | Increase AV volumes to 5% of new car sales | 15%–20% sales are AV; lower-price models | Follow the leaders |
|   | Strategy: Tech Cos. Waymo, Tesla etc. | Level 3 and level 4 for TaaS in small quantity; level 5 trials; own OS and AI | Build on level-5 TaaS success; license autopilot; build trust | Deliver level 5 to private owners; more OEM partners | Form joint partnership with OEMs; Push OS & autopilot as defacto std. | Joint Tech plus OEM co. as industry AV leader |
| 7. | Connected AVs | Early discussion; no standards | Some early standards and prototypes | Systems integration using standards | Connected AVs prototype trials | Connected vehicles mature |

| | | Phase 1 (2017–2021) | Phase 2 (2021–2025) | Phase 3 (2025–2030) | Phase 4 (2030–2040) | Phase 5 (2040–2050) |
|---|---|---|---|---|---|---|
| 8 | Electric AVs (Bloomberg)57 | Less than 5% of cars | EVs pick up | 8-24% | 24-43% | Majority EVs Minority ICE |
| 9. | Smarter AVs | OEMs not offering Apple CarPlay or Android Auto | Consumers resist OEM HMI, want unified HMI with infotainment | Smart AVs show up, connected car appears | OEMs adopt standard Smart framework | Smart car framework gets defacto standard |
| 10. | ADAS | Consumers begin buying ADAS | Consumers prefer ADAS to traditional driving | More ADAS-featured cars than AVs | AVs overtake ADAS-featured cars | Dramatically more AVs than ADAS-only cars |
| 11 | Sensors, LiDAR, Cameras | First generation expensive | Prices coming down; better configurations | Second-generation sensors | Evolution continues | Commodity item |
| 12 | Infrastructure Upgrade | No upgraded expected | Some smart cities will upgrade | More cities will upgrade | Trend will continue | Most major cities will upgrade |
| 13 | AV & Non-AV Mix on Road | AVs only for trials (Less than 100K) | AVs reach 500K in 2025 (Only 0.2%) | AVs reach 5 million mark (Only 2%) | AVs Reach 20 million mark (10%) | Reach 80 million (50%) |
| 14 | Benefits Realized: 90% Fewer Accidents | No benefit in this period | More accidents initially in hybrid setting; Geo-fenced trials show potential for reduction | Data for lower accidents available | Humans learn to co-exist with AVs; some benefits in smart cities | May start realizing measurable reduction in accidents |
| 15 | Benefits Realized: Fewer Cars | No benefit; congestion getting worse | No benefit, test data from smart cities may help | Data from smart cities for future planning | Smarter cities may see benefit, but less than forecast | Smarter cities may see benefit; still less than forecast |
| 16 | Benefits Realized: Less Congestion | No reduction | No reduction | No major reduction (AVs will drive more slowly) | Smart cities with new infrastructure see some benefits | Tangible reduction; much less than forecast |
| 17 | Benefits Realized: Reduced Insurance | Some signs already (reduction for ADAS-featured vehicles) | Insurance companies announce reduced rates, if OEM liability | Premiums go down for AVs, up for non-AV's | Trend continues | Premiums level off on real safety data |

|    |             | Phase 1 (2017–2021) | Phase 2 (2021–2025) | Phase 3 (2025–2030) | Phase 4 (2030–2040) | Phase 5 (2040–2050) |
|----|-------------|---------------------|---------------------|---------------------|---------------------|---------------------|
| 18 | Regulations | Policy guidance only, AV levels defined | More details, third-party certification prescribed | Regulation revised based on experience | Better regulations | Regulations settle down |

*Table 14.1: Five phases of AV evolution*

## The AV Adoption Path

The following trends will define the path for AV development, manufacturing, marketing and adoption:

**Evolutionary path to an AV revolution (supply side):** While some analysts and business innovators have created the impression that AVs will cause a lot of disruption, I believe this disruption will be slow, gradual and incremental. If somebody looked at where we are today, slept for 40 years and woke up in 2058, they might very well say that it was a revolution. But if you were awake during those 40 years and witnessed the gradual progress and painful development with many setbacks, workarounds and redesigns, you'd likely use a different word to describe that change. To me (and to the Oxford dictionary), the word "revolution" connotes a catastrophic event that introduces extensive changes suddenly. I don't see a revolution here. There are enough counter forces that will turn revolution into evolution. I don't see strong established OEMs disappearing from the auto landscape, either. They'll adapt and fight a formidable battle with their challengers. Those OEMs and tier-one suppliers, who are not nimble, who don't innovate and change with technology trends, will disappear.

**Consumers hold the key (demand side):** While AV development has attracted a lot of attention, I think that the real key to adoption lies with the consumer who so far lacks trust in AVs. I talked about consumer trust issues in chapters 9 and 12. Some AV enthusiasts and promoters would have OEMs believe that if they build AVs, consumers will swamp the dealerships or ditch their cars and start exclusively using AVs for ride-hailing or ride-sharing, TaaS-style. It's not going to happen that swiftly. Except for a few early adopters, consumers are going to wait till they're convinced that AVs are 100% safe and that the price is worth the benefits. They're not going to give up control. Waymo and the leading auto OEMs have a huge marketing effort ahead of them to win the trust of consumers, especially the baby boomers, older millennials. Y and Z generations will adopt but gradually. It will not be a tide of adoption like it was for smart phones.

**Long elapsed period (30–40 years):** I created an Excel spreadsheet to project how the mix of AVs and non-AVs on the road might change over the next 30 years. It shows that it will take upwards of 30 years for the majority (more than 50%) of new-car sales to be level-4 or level-5 AVs in the developed world.

1. **The future is hybrid, for at least 40 years:** I use the term "hybrid" here in a generic sense to mean the co-existence of non-AVs (going down over time) and AVs (going up over time). My simple calculations based on an Excel spreadsheet show that we'll have a hybrid situation—with both AVs and non-AVs on public roads—for at least 40 years after the first fully autonomous AV drives human occupants as passengers. If that happens in 2025, that means it'll be at least 2065 when 50% of the vehicles on the road are AVs.

2. **Anticipated benefits of AVs will only be realized in the final phases:** I project that initially (in phases 1 and 2), there may be *increased* accidents and congestion because of the confusion between human drivers and robo drivers. Gradually this confusion will decrease and human drivers will learn to co-exist with AVs. ADAS features will help to reduce accidents. So it might be wash in terms of accident statistics. In phase 3, traffic will become more organized and will follow the revised rules of the AV regime. The promised benefits will start trickling in. But most of the benefits will be realized only when the majority of vehicles on the roads are AVs. That means we'll have to wait till 2050–2060.

   We've seen that legislators, regulators and DMV officials don't often make long-term planning decisions in favor of future generations (many legislators are singularly focused on the next election). If expedited adoption of AVs forces them to adopt visionary thinking, it'll be great for our grandchildren.

3. **AV adoption won't be like the internet and smartphone revolutions**: Some innovators compare the AV revolution with two major technology innovations in recent past—the internet and smartphones. But both of these were brand-new, and the underlying technology in both cases was well-proven. Widespread adoption was simply a matter of implementation on a massive scale across the globe. Customers were ready and eager—in fact, they were lining up to buy. AVs, on the other hand, are replacing an existing, known technology—albeit one that causes congestion and deaths—so adoption rates will be slower.

## 14.6 The Five Periods of AV Evolution

### 14.6.1 Phase 1 (2018–2021)

(Confidence level of prediction = 80%)

This is where we are now. AV trials, development and refinements are in the works. Market maneuvers (acquisitions and partnerships) are taking place. Scientists and researchers are still trying to find solutions to the unsolved bits of the AV problem. Hardware component providers are trying to pack more functionality into cheaper silicon with smaller footprints and more efficient power requirements. Robo-autopilots are being coded. OEMS and technology challengers are doing trials in those areas where the infrastructure managers are willing to let them do so with minimal restrictions. Timelines are considered more important than the mission-critical nature of the task. Regulators are letting the industry make its own rules. No standards of certification i.e. criterion of passing and failing have been defined and current laws govern the driving on the road. Public relations efforts by AV proponents are moving at full speed — expectations are high.

And yet this is the most important period in our journey because we're laying the foundation for AV introduction and adoption. We must get it right. I expect that by the end of this period, suppliers will have gained enough insight into the limits of the current technology to start offering something realistic that will keep the dream alive in the eyes of the consumers.

Here's what I expect to see by 2021:

- Major AV suppliers would start offering TaaS offering for fee either directly or through arrangements with Uber/Lyft. There would be restrictions on the offering but it will validate their claims that they have met their announced timelines.
- TaaS is offered in geo-fenced areas of several AV-friendly cities using level-4 and level-5 AVs at prices that are more attractive than the current Uber/Lyft business model.. The objective will be threefold—gain consumer trust in the technology, test the technology for further debugging of the robo-software and experiment with backend support systems. The TaaS providers are not ready to wean off from human drivers in AVs.
- DOT has defined a mechanism for designating areas where AVs can operate without human drivers.
- Uber/Lyft and other TaaS providers are offering autonomous ride hailing services at $1.00 per mile. SAE level-5 services are also offered in university campuses, hospitals, airports, small suburban municipalities, etc., using Navya-like AVs at prices lower than current Uber/Lyft prices that are around $2.50 per mile.
- SAE level 4 AVs (Tesla like) are common on designated interstate highways with human drivers encouraged to take over as the cars exit the highway.
- State DMV agencies have started designating road segments that are AV-ready. Routing and planning functions exist in autopilot software is capable of figuring out whether there exists an AV-designated route from point A to point B.
- There's more clarity from DOT-like federal agencies regarding national AV regulations. The United States will continue to take the lead, with other advanced countries in Europe following US guidelines but with stricter regulations.
- Independent testing and certification labs, like the American Center of Mobility in Michigan, have emerged.

## 14.6.2 Phase 2 (2021–2025)

*(Confidence level of prediction = 60%)*

The march to our AV destination continues. The path is now clearer—suppliers recognize what needs to be done and where the roadblocks are. I see the following terrain through my crystal ball:

- The TaaS business model is becoming better defined as another channel to meet transportation needs. After trials with different pricing concepts and promotions, the industry recognizes its limits (TaaS will wipe out the taxi industry, but it's not going to replace our obsession with car ownership so fast). The business is there but profit margins are low. Human drivers are still needed in congested cities because consumers are not yet comfortable with the level-5 AV technology. At the end of this period, there are TaaS trials without human drivers on protected city-street routes with remote TaaS chasers (human drivers who can take over AV in trouble) nearby, available at the touch of a button or a robo-911-call away.
- SAE level-5 AVs are moving further along to their final destination but only in controlled settings.
- The BOM (Bill of Material) for components to build a level-5 AV is still high in 2025—even though it has come down since 2018. The quantities are still not high and mass production has not begun because AVs are still being built in small numbers, so price-differential between AVs and non-AVs is still high. Some experts are quoting a figure in the $15,000 range, required if the autos OEMs are to make a decent margin on AVs and get a reasonable return on their R&D investment.
- By 2025, OEMs are ready to sell AVs to private buyers with some restrictions. Regulators continue to restrict these to highways only. It's not clear whether cities will allow level-5 AVs to drive on public roads.
- Insurance companies have reduced insurance prices for AVs, as long as manufacturers assume a certain amount of liability.
- Technology challengers have conceded the manufacturing battle to established OEMs and formed joint partnerships.

## 14.6.3 Phase 3 (2025–2030)

*(Confidence level of prediction = 50%)*

In this period, AVs are facing their first real workout on public roads with real consumers—TaaS passengers (levels 4-5) or private buyers (levels 3–4 only). OEMs have the following vehicles and services in their show rooms:

a) Non-AVs: Electric and ICE power with no active driving automation for the less affluent consumers—who still form a large majority of the consumer base
b) Non-AVs with advanced ADAS features under a human driver's control (levels 2–4)

c) AVs (electric and ICE): Levels 3 and 4—primarily in luxury/semi-luxury category
d) TaaS bundles in the form of flexible AV leases e.g. 200 trips per month with 1500 mile ceiling or 1000 passenger mile bundle in any period.

The majority of auto sales continue to be in the first two categories, with significant sales in the second category. AVs in small numbers (hundreds of thousand units, not millions of units per year) are being sold to TaaS providers as corporate sales. TaaS trials without human drivers have started in controlled environments in on some public roads certified as "AV-ready" by DMV transport staff.

Other trends that I predict for this period:

- Fully autonomous vehicles have been tested on many public roads, even in the downtown core of many major cities in the developed world, including China. Every year, human intervention in these trials has been decreasing. Level-5 AVs from technology suppliers like Waymo and a few leading OEMs still do not meet "six-sigma" quality-control standards (two to three human interventions per million miles driven) in mixed traffic, but they can do so in highway traffic. The DOT has now specified human-intervention standards. Suppliers have now indicated that they will sell level-5 AVs to private buyers only after 2030.
- There's a debate in the industry about taking pedals out of level-5 AVs. The DOT does not want to pass regulations in this regard. The industry is now saying that some accidents may still occur with level-5 AVs but that there will be significantly fewer accidents and casualties than with non-AVs without active automation.
- The Chinese have started allowing level-5 AVs (both in TaaS mode and private vehicles) on public roads. Consumer acceptance is relatively low.
- Several cities in other countries have taken the lead from Singapore's TaaS experiment and have started public discussions about banning private vehicles in downtown cores.
- Automakers now understand the mission-critical nature of fully autonomous cars and are discussing the need and economic case for redundancy design for critical components for level-5 AVs.
- The number of vehicles on the roads has not decreased during this period, nor has the congestion in cities. But there is still hope, because the Singapore experience shows it can be achieved in a controlled setting.
- City organizations have set up task forces to upgrade road infrastructure to enable AVs. Industry lobby groups are asking that city transport departments designate special lanes for AVs. The public opposes that idea and debate continues.
- The public, especially non-AV drivers, is opposed to platooning (vehicles driving in formation) being allowed on highways because they hog large portions of the highway.
- Workers' unions have started courting their legislators about the loss of jobs due to automation, especially in the freight industry.

- The regulators like the DOT have issued more detailed guidelines, including testing and certification requirements.

### 14.6.4 Phase 4 (2030–2040)

(Confidence level of prediction = 35%)

Now the AV adoption race is on at an accelerated pace. Suppliers are no longer in the trial mode of the previous decade. They've spent a lot of VC money or their own capital to turn AVs into a profitable business, but they have not made much headway so far; the return on the AV investment remains a future objective.

Over the past decade, the industry has tried to gain consumer trust through Drive Me–type trials and subsidized TaaS fares, with and without human drivers, and has provided human-intervention and accident statistics that show that these services have a better safety track record than non-AV vehicles.

I forecast the following trends during this decade:

- The number of new AVs (levels 4 and 5) on the road is around half a million, and growing at a CAGR of 15% for the first five years and at 25% during the next five. The majority of these are used for TaaS. However, by the beginning of the decade, private sales of AVs have started, predominantly to the Y and Z generations. Second-car ownership has dropped significantly—people are using TaaS for the short trips for which they used their second car. Consumers are willing to accept sharing TaaS for office trips but not for family trips.
- While the emphasis is still on the self-driving aspect of AVs, customers do want, and the industry has promised an integrated, well-connected, smarter and greener AV with an intuitive HMI and a voice interface using a conversational NLP. Only by the end of the decade do we realize the promised AV Plus described above..
- There's a marginal decrease in the number of cars on the road. However, by 2040, smart cities see a measurable decrease.
- Consumers driving less than 5,000 miles annually are switching to TaaS. However, TaaS has not taken off among rest of consumers as projected by industry evangelists in 2018 because TaaS fares for single rides are higher than the operating cost of a privately owned vehicle. One major assumption of the TaaS model is sharing among multiple passengers—a practice that consumers do not like because of the inconvenience, extra trip time and lack of privacy.
- Private ownership of cars has dropped slightly because of two-owner families downgrading to one car plus TaaS but congestion on the road has not decreased because of natural increase in population base.

### 14.6.5 Phase 5 (2040–2050)

(Confidence level of prediction = 25%)

This is the period when AV will reach a tipping point—when there are more AVs sold than non-AVs among new car sales.

AV trends during this period:

- The prices of AV computer hardware and software components have dropped significantly.
- Consumers (generations Y and Z predominantly) have become comfortable with the concept of self-driving cars.
- OEMs are offering a full range of AVs (levels 3–5), non-AVs and TaaS services through their dealership network or affiliated organizations. OEMs are offering AVs for private ownership. OEMs still sell more non-AVs than AVs. Less affluent consumers in all generations do not find AVs or TaaS affordable.
- AVs are now mostly connected, electric, smart, infotainment-centric—AV Plus indeed.
- Many cities have started upgrading their road infrastructure. Cities are designating road sections, intersections and highway exits "AV-capable." Route-planning software has been upgraded to recognize this attribute when calculating an AV route.
- Robo-autopilot is able to mimic the human driver quite well, although it still faces situations it cannot handle. This happens very infrequently. A safe fallback procedure has been designed and approved by DOT for when the robo-autopilot has to stop on the road. Multi-faceted AV tow-away procedures are in place.
- AVs and non-AVs continue to share the road, and there are still more non-AVs than AVs.
- 10% of all new light vehicle sales (20 million in North America, including Canada) are AVs (levels 3 to 5) at the beginning of the period, accelerating to 30% by the end of 2050. This represents CAGR of about 15%.
- OEMs have started benefiting from the supply of higher-margin apps for their vehicles.
- A typical family relies on a combination of non-AVs, AVs, TaaS and public transportation to meet their transportation needs. Private ownership of non-AV cars has decreased but still exists.
- Regulators have made some accident-avoidance ADAS features mandatory on all new cars just as air bags and seat belts became standard equipment decades ago in North America.
- Insurance premiums have dropped. Actuaries are still waiting for reliable accident data to help reduce them further.
- The majority of the regulatory, legal, insurance and ethics issues have been resolved.

## Summary

*This chapter emphasizes that the evolution of AV will be slow, gradual and incremental. Consumers will determine the adoption rate, and they have options. Self-driving is exciting but not compelling. Benefits will accrue but only after 30–50 years and not to the extent forecast by today's proponents.*

*The predictions are based on lots of assumptions and have lower confidence level as we project ourselves further in time horizon. That is reality.*

---

### Citations for External References

[43] What AV Executives say about AV future? – https://www.driverless-future.com

[44] IEEE Says Audi and Nvidia Join hands -- https://spectrum.ieee.org/cars-that-think/transportation/self-driving/nvidia-ceo-announces

[45] http://europe.autonews.com/article/20170917/COPY/309179994/volkswagen-readies-its-robo-fleets

[46] Ford CEO says about AV timeline – https://www.reuters.com/article/us-ford-autonomous-idUSKCN10R1G1

[47] GM AV Future – https://www.cnbc.com/2017/11/30/gm-to-launch-autonomous-cars-in-big-cities-sometime-in-2019.html

[48] Andrew Ng – https://www.quora.com/Andrew-Ng-When-will-self-driving-cars-be-on-roads?no_redirect=1

[49] Driverless Taxi firm Nutomy – https://www.yahoo.com/news/driverless-taxi-firm-eyes-operations-10-cities-2020-142503529.html

[50] Intel CTO – https://www.yahoo.com/news/driverless-taxi-firm-eyes-operations-10-cities-2020-142503529.html

[51] Look Ma No Hands – https://www.ieee.org/about/news/2012/5september-2-2012.html

[52] Andrew Moore in Forbes – https://www.forbes.com/sites/alanohnsman/2016/09/28/warning-driverless-cars-are-farther-than-they-appear/#69bde98f6427

[53] MIT Technology Review – https://www.technologyreview.com/s/602210/prepare-to-be-underwhelmed-by-2021s-autonomous-cars/

[54] MIT Technology Review – https://www.technologyreview.com/s/602210/prepare-to-be-underwhelmed-by-2021s-autonomous-cars/

[55] Deloitte Survey of Consumers -- https://www2.deloitte.com/content/dam/Deloitte/us/Documents/manufacturing/us-manufacturing-consumer-opinions-on-advanced-vehicle-technology.pdf

[56] Deloitte Survey of Consumers -- https://www2.deloitte.com/us/en/pages/manufacturing/articles/automotive-trends-millennials-consumer-study.html

[57] Bloomberg Electric Car Report https://data.bloomberglp.com/bnef/sites/14/2017/07/BNEF_EVO_2017_ExecutiveSummary.pdf

# Chapter 15

# The Way Forward for AV Plus

*An emerging technology that has no precedent must depend on trial and error methodology by the proponents. With so many challenges to overcome, especially consumer attitude, it is difficult to chart a clear path to the destination. Yet there are steps that the industry and external stakeholders can take to minimize the delays on the route. I intend to discuss those steps that different players can take in my closing chapter.*

The main thesis of my book is that it will take a while for the AV evolution to happen—much longer than the proponents would have us believe. Immature technology, uncertainty, lack of trust in a mission-critical technology, a self-centered attitude by individuals toward societal benefits, consumer resistance to change, lack of readiness of city infrastructure and flawed TaaS economic models all contribute to my conservative outlook. I don't think lack of regulations is the main reason. Predicting the future is a difficult endeavor even for experts; I've ventured to do it but only by couching my predictions in confidence levels that go down from 80% to 25% as I predict things further and further into the future. Is there a way to expedite the AV evolution? I'll offer some advice that could allow the industry to move forward a little more quickly.

**AV Plus**—safer, smarter, connected, greener, cyber-protected and equipped with a conversational human interface—is on its way, out of the labs and onto trials on protected public roads. There are millions of vehicles already equipped with ADAS features throughout the world now—some active and some passive. These ADAS-equipped vehicles can perform some of the functions that AVs will do, but currently a human driver must still be involved. Some of these ADAS features may become standard equipment in future vehicles—whether voluntarily by auto OEMs or mandated by regulators. ADAS-equipped cars, as a first step toward AVs, are reducing accidents and making us safer on the road, although not to the same extent as we can expect from fully autonomous vehicles in the future.

The next major step in this journey is to replace the human driver with a robo-autopilot. This is where some of the brightest scientists, engineers and innovators in the Silicon Valley and elsewhere are working today. They're using cutting-edge AI techniques and powerful supercomputers. **The task of equipping a vehicle with an artificial brain that can replace the human brain in driving safely is not easy.** Some AI experts equate it to President Kennedy's moon-shot project. In my view, it'll take more research and development, more engineering design, more trials, more independent testing, more fresh thinking, more consumer trust, more design revisions, more twists and turns but most importantly more perseverance on the part of the innovators.

Who's leading the charge? Twelve years after the 2005 DARPA Grand Challenge and eight years after Google started its self-driving project in 2009, all major OEMs and a few technology companies are in charge. That's our competitive free-enterprise system in action. Established OEMs are trying hard to protect their turf as the technology companies challenge them. The two groups are working in parallel as well as crisscrossing each other's paths. The incumbents have their pride and are not yet ready to embrace the challengers, who they see as threats with their market caps and investment-community backing. On the other hand, the technology challengers must realize that it takes a certain skillset, knowledge base, business network and manufacturing infrastructure to serve the needs of millions of vehicle owners. The OEMs are partnering with specialized technology companies hoping to compete vigorously in the new landscape. Incumbents feel both threatened by their own weaknesses and smug in their own apparent strengths. They're not used to business models involving high-margin consumer apps and advertising revenue. Taking advantage of new revenue streams in a digitized and connected AV world is certainly not their cup of tea. But they're going to have to learn and take part in order to survive.

The record of auto OEMs is not stellar when it comes to digital transformation or cellular communications in cars. GM's OnStar and Ford's Sync system were both relative failures for many years, although they have been revitalized now and are showing some profitability. After I had a free OnStar subscription for the first year in my 2005 Buick Allure, I never renewed the service because the monthly fee simply didn't justify the benefits, especially since I always had a cellphone with me that was so much more versatile. OnStar didn't initially offer Bluetooth connectivity to the owner's cellular phone—that came much later.

Further, the car companies have done a relatively poor job on the whole in terms of a navigation-system interface. They forced buyers to pay a fairly high price to use their car's navigation system, and they met with a certain amount of reluctance and dissatisfaction. Consumers preferred the simple interfaces of inexpensive TomTom or Garmin standalone units to the cumbersome, hard-to-use vehicle navigation system of the car. When OEMs wanted to charge $300 to $500 to update their systems' map databases (which we got for free from Garmin or TomTom through a very simple interface that any computer-literate person could do), consumers considered it a rip-off. Compare that with Tesla's OTA software updates. The final nail was driven into the car companies' navigation coffins when real-time maps appeared on smartphones **for free**, especially with Google's Waze app providing real-time traffic congestion information as well. Why would you use an OEM's navigation database that isn't current to begin with and has a clumsy user interface to boot?

OEMs' reluctance to accept the superiority of the smartphone for communication or navigation did not end there. For the past three years, OEMs have not been willingly providing support for Apple CarPlay and Android Auto in their new models. It's not difficult to do it technically; OEMs simply feel threatened by the idea of giving control of the car's console to technology challengers. The establishment has not learned; they are still resisting the inevitable. They simply can't create a superior competitive offering that connects into customers' smartphones. There are two to three billion (and growing) smartphones with iPhone and Android interfaces and hundreds of thousands of apps—most of which are free.

Nobody doubts any longer that we will one day ride in autonomous vehicles. We all agree that the transportation future looks *AVy* (a new adjective I just coined). Of course, there are many questions about our *AVy* future. The first question I have is whether older baby boomers, like me, will own one in their lifetime. My generation *has* started buying ADAS-equipped cars. I own a 2016 Mercedes C300 with Intelligent Drive—a low level-2 AV that gives me a glimpse of the future. I can speculate that my Generation X children will consider owning an AV and will perhaps downgrade from two cars to one, supplementing one of their cars with ride-hailing TaaS for their remaining transportation requirements. As I look further into the future, I can visualize my grandchildren willingly adopting AVs without much of the fear or lack of trust that my generation has.

I've already speculated, guesstimated and predicted in Chapter 14 when AVs will show up on public roads as ride-hailing services or in dealer show rooms. Here in this final chapter, I want to provide some guidance to the various AV stakeholders that I think could help expedite the arrival of AVs. I project that the industry can accelerate the adoption of AVs by as much as five or even 10 years if they follow this advice. My fear is that they won't.

## Should the Industry Follow an Incremental Path or Go for the Moon Shot?

Established auto OEMs have been around for a long time—some more than hundred years. They are generally traditional in their strategies and careful with their money. They use proven technology architectures and business models, and they stress an incremental path to innovation. They generally have small think tanks that watch others' innovations to see what might apply to their industry, and some of them, like Daimler, GM and Ford, have participated in the autonomous vehicle research, trials and competitions conducted by the universities and DARPA. They certainly never thought that the AV train would leave the platform without them. However, technology challengers like Waymo/Google and Tesla have proved them wrong. When the challengers excited the imaginations of the media and the public through self-driving trials, OEMs got scared and reacted defensively to protect their turf, hurrying to catch the train and meet the challengers at the next station. But their preferred strategy is an incremental upgrade from non-AVs without ADAS features to non-AVs with ADAS features, leaving the active control with the human driver. They plan to introduce AVs—level 3 to level 4 and ultimately level 5—separately, in parallel.

The technology challengers, in contrast, have upped the ante by rejecting the incremental-upgrade strategy and racing full-speed down the track toward fully autonomous cars, setting SAE level 5 as the target with a delivery timeline of 2019–2021. They use two key assertions to support this strategy. The first is that car accidents/casualties will be reduced and safety will increase if we have fully autonomous vehicles on our roads exclusively, with no human drivers. In a perfect world of highly reliable AVs and mass adoption, this is an entirely valid assertion. The second is that we cannot expect a human driver to take control very quickly when the robo-driver needs them to. The human driver needs time to figure out what's going on and then take control. Therefore a robo-driver that relies on human control in emergencies is not feasible.

There's a great debate as to which strategy is better—incremental or moon shot. **I think both are valid strategies and that they're complementary to each other.** Let me explain why I think so. My opinion is that the industry should follow both strategies in parallel and that industry influencers (consumer bodies, the media, consultants and others) should encourage auto manufacturers to pursue both. In fact, some OEMs like GM are doing exactly that—they have announced that they'll have a level-5 car ready in 2019 along with a level 3 and a level 4. They want the marketplace and consumers to vote on which way to go.

From migration, consumer-acceptance and adoption points of view, the incremental strategy of the established OEMs is superior and more pragmatic. I believe that the incremental strategy will succeed in the short and medium terms (phases 1 to 3 in my definition in chapter 14; see Chapter 14). The majority of the consumer community will prefer the incremental strategy; that's how consumers, by and large adopt technology. They want to be comfortable with step 1 before going to step 2. We might be willing to ride a new technology wave in a single step *if* it gave us something that we never had, like in the smartphone revolution, which took only a decade to reach 75% of our population. But with AVs, the utility function gives us something we already have—the ability to get from point A to point B. The incremental strategy will create less disruption and will give regulators more time to respond and cities more time to upgrade their infrastructure. It's almost impossible to expect that the entire vehicle base could be converted to level-5 AVs in a short period. As I said in Chapter 14, I believe this adoption will take two to three decades, and even then there will still be a huge percentage of non-AVs on the roads. A large percentage of our population (across all generations) would prefer an incremental approach—they want to ease into the fully autonomous regime gradually. The grandchildren of the baby boomers (Generation Z) will be among the first to adopt level 5s, and it'll be many years before level 5s are the predominant cars on the road. I believe the Waymo moon-shot strategy to level-5 autonomy without pedals and steering is technically risky and impractical from a business-model perspective.

The reason I'm suggesting that we should encourage avant-garde innovators like Waymo and small teams in the OEM establishment to keep plugging away at level-5 AVs is that the overall rate of progress toward the final goal could slow down if these ambitious companies aren't pursuing their focused, single-minded strategy. Regulators should continue to support these companies in testing level-5 AVs on public roads, although they should certainly be imposing high standards of compliance through testing and certification. In this context, the DOT and NHTSA should work to establish problem-resolution expectations for AVs, to avoid having occupants get stranded when problems arise (as they are bound to). This will be a win-win policy because the "moon-shot" group will benefit from a first-mover advantage and can then license their robo autopilot software to the "slow-and-steady" group. The slow-and-steady group, made up of mostly traditional OEMs, will continue to have an advantage in terms of assembly-line infrastructure. This will create inter-dependency between the two groups—something that consumers will benefit from.

|  | Incremental Strategy (Non-AV → ADAS → Level-3/4 AVs → Level-5 AVs) | Direct to Fully Autonomous Level-5 Vehicles |
|---|---|---|
| Who is pursuing it seriously? | • Established auto OEMs | • Technology challengers such as Waymo |
| Pros | • Gradual and natural adoption by consumer<br>• Co-existence of AVs and non-AVs supported for a longer time<br>• Cities prefer this strategy<br>• AVs must offer value to be adopted | • Consumers see private ownership of level-5 AVs earlier<br>• Benefits, such as reduced accidents/casualties, realized sooner |
| Cons | • Less committed to level 5—more focus on ADAS, L-3 and L4 AVs<br>• Level-5 announcements for "me too" public relations primarily | • Difficult to achieve in short-term<br>• Consumer trust lacking<br>• Consumers stranded if problems |

## 15.1 What Should AV Stakeholders Do?

There are several broad issues that should be addressed on an industry-wide basis, ideally by an industry association. While there are a number of auto-industry associations, I feel that there should be a specific AV-industry group. One organizational model that could be followed is that of the Cellular Telecommunications Industry Association (CTIA). CTIA addresses industry-wide issues and deals with regulators on regular basis.

An AV industry group or association should be formed with the following objectives:

- Interface with legislators and regulators on common industry problems and issues
- Interface with the insurance industry (with consumer representation)
- Come up with acceptable ethical scenarios for robo autopilot software
- Interface with the NHTSA and SAE (and other such bodies) regarding testing and certification issues
- Clarify communications standards issues such as DSRC and 5G confusion for V2V, V2I and V2X communication
- Discuss and debate user-interface standards on infotainment and smartphone interfaces
- Provide public education and build trust among consumers

## 15.2 What Should Established OEMs Do?

There are a number of steps that established OEMs can take to help speed up the AV evolution. I propose the following measures.

## 15.2.1 Cultural Shift

The auto industry is old and established; it's also very traditional in many respects. This aspect shows up in OEMs' boards, senior executive teams and employee ranks. For decades, they have met the challenges of the past and created outstanding product-design teams, deep systems-integration expertise, efficient supply chains and fine-tuned, automated assembly lines. But the OEM of the future will need to be different. OEMs' past success is being challenged by the emerging future of AVs. The future calls for OEMs to be innovative, nimble and disruptive. The stock market sent a clear signal in 2017 when it put Tesla's market valuation at more than Ford's. OEMs need to revitalize themselves. Even though it has served the industry well, there is a need to change the fundamental culture of the auto industry at all levels—at the board level, at the technology level, at the operational level and at the business-model level. Everything has to be reprogrammed. OEMs need new, young, techno-savvy blood, and a hybrid business culture that values the past but emphasizes the future more. **The change must be pervasive, starting right at the top: in the board.**

In an article entitled "The Auto Industry Won't Create the Future" in _Wired magazine_[58], author David Pakman summarizes the cultural differences between the technology companies and the established auto companies and looks at what needs to change for the OEMs to flourish in an AV world. I generally agree with his points. He suggests that OEMs need to adjust in a number of areas, including the following (in which I've also added some suggestions of my own).

**Innovation:** Industrial age industries like auto must introduce more technological innovation than they have in the past. The innovation must be continuous, substantial (not just changing the trim or the shape of the car) and strategic. Innovation must come not just from tier-one suppliers but also from within auto companies (as happens at Apple and Google).

**Electric, not ICE:** The auto industry had been dragging its feet for 20 years until Tesla proved that long-range electric cars could be built by a vertically integrated company, and built beautifully, with fewer parts, superior reliability and quicker rejuvenation frequency (through OTA updates of auto pilot). OEMs' excuses were proven wrong. The only thing left to discuss is the pace at which we will move toward electric AVs.

**Use of software/firmware**: As vehicle components get more and more computer-controlled, there's an opportunity to not only get maintenance information but to also change vehicle behavior by sending changes to the software and firmware over the air rather than through a dealer network. This is something that technology challengers can do well. OEMs will need extensive computer and management systems to do it for a large base of several million vehicles.

**Give customers something free and get something valuable in return**: Technology challengers (like Google) and the auto establishment have diametrically opposite marketing philosophies. Google cultivates and nurtures a new market by giving freebies to millions of consumers. Free internet search, free cloud storage, free photograph catalogs, free maps, free map updates, free Waze navigation application (the best in the business), free email service, free smartphone apps and lots more are given to anybody and everybody, even the competition. You can opt in and opt out at any time. Google does this because they want you to share something with them—your data.

When did you last get anything free from your car company or your dealer, even though you bought your second most expensive asset from them? They want $200–500 plus a visit to the dealer just to update your navigation maps. In order to have car owners subscribe to their advertising database and part with personal info, OEMs have to use similar techniques that Google/Facebook/WhatsApp use to entice customers – give something free.

**Direct relationship with customers**: While OEMs rely on dealer networks to sell their products and interface with their buyers, technology companies build direct relationships with their customers. Tesla has already adopted that arrangement by using company-owned showrooms. Look at Apple stores—people flock to them. Better margins plus ownership of the customer relationship show up in better returns and stock valuations.

### Will There Be or Should There Be a Merger?

If the boards could be persuaded and if personal pride of key management team members could be set aside, it might serve organizations and consumers better if an OEM and a technology challenger decided to merge with one another. Their different cultures and skillsets just might make for a strong combination. Perhaps a subsidiary owned by two companies could be created. What I see so far are new organizations that are dominated by one party or another. I am sure it will happen under the nodding and nudging of VCs.

### 15.2.2 TaaS First, Private Ownership Later

There are some AV enthusiasts who suggest that private vehicle ownership should be discouraged—some people are even proposing a ban on private ownership in congested areas like downtown Manhattan. Instead, they say, all transportation should be offered as a service (for a fee) that can be called from a smartphone. The city of Singapore is going to try such an experiment. While banning cars in designated areas is a regulatory issue, AV suppliers can only delay this private ownership of AVs through indirect measures. Established OEMs see TaaS as a business opportunity that can shore up their declining revenue streams in the future (on the assumption that TaaS will have higher profit margins) as the number of cars on the road decreases. Perhaps it's this motivation that is encouraging the industry's "TaaS first, private ownership later" strategy. If a TaaS provider is the AV manufacturer, then that manufacturer had better make that AV as reliable as possible, especially if they're taking on the liability.

Such a strategy is a win-win for both the industry and consumers. I'm in favor of TaaS as the first step toward winning the public's trust. Private ownership can follow.

### 15.2.3 Offer AVs in Three Categories—Medium, Semi-Luxury and Luxury

It makes sense that AV manufacturers would start by offering luxury versions of their AVs. The physical design will change as we get close to the 2021–25 timeframe but the price category may not change. Affluent customers are less price-conscious so long as they like a concept. Early, affluent adopters may buy AVs because they want to join the leagues of technology adopters and because they want to show off a new toy. However, early adopters won't make a big dent in the adoption figures. Serious adoption will happen when there are AVs in all three categories—the medium-price category, the semi-luxury category and the luxury category. This will cover a much wider consumer base across all generations—baby boomers, Generation X, millennials and Generation Z. Whether marketing strategists in the auto companies will think it makes business sense to do so is a different matter; the common strategy is to offer superior features in the luxury category first and slowly migrate them to the lower-priced categories.

TaaS service providers will have to offer different AV configurations with multiple fare categories (from a private luxury class to a shared budget class) because the customer base will vary from affluent to frugal. Uber is doing this already.

### 15.2.4 Keep the Price Premium for AVs Reasonable

It's the million-dollar question: what will be the price premium between non-AVs and AVs of the same class and features? Only the marketing and business planners in OEM business-development groups may know what's coming; I don't have access to that information. But I'll throw around some numbers for discussion purposes. Customer surveys by Deloitte show consumers will only accept a price differential in the $1,500 range, but I don't think this is a reasonable number. I believe that the starting premium in the luxury category will be $10,000 to $15,000.

Consumers in the intermediate category will find a $15,000 premium excessive. The obvious result will be that they will postpone buying an AV and instead will buy a non-AV or an ADAS-featured vehicle when they're ready to replace their current car.

OEMs know the pricing game very well and will conduct extensive consumer surveys before deciding what premium to charge in the first year. My crystal ball suggests there will be slow adoption as a result of high price premium. Remember, consumers have a choice: they can still buy a non-AV, which provides the same function and almost the same level of safety. It'll be a long time before individual consumers with good driving habits are convinced that AVs with robo-autopilot are safer than ADAS-featured non-AVs with human drivers.

### 15.2.5 Don't Develop AV Apps—Just Host Them

The major consulting houses have a lot of advice about how the auto OEMs can develop new revenue streams. One such stream is in hosting apps on OEM servers. This is perfectly valid advice. But there could also be a temptation by OEMs to start *developing* auto infotainment apps or auto versions of smartphone apps. Based on the experience of the smartphone industry, where Apple and Google purchase price, OEMs should steer clear of developing their own. Apple and Google don't develop the majority of apps; they simply provide a development framework, tools and a platform for customers to install the apps to their devices. OEMs should adopt a similar approach. They can get significant margins from the purchase of movies, videos and other content streamed from content providers. OEMs can also expect a significant amount of advertising revenue because they'll have a captive audience.

### 15.2.6 Don't Fight the Telecom Carriers—Become MVNOs Instead

Large OEMs can adopt the mobile virtual network operator (MVNO – a concept that allows OEMs to become wholesalers of communication services) business model and use it as a revenue-generation tool. OEMs will need 5G cellular services for updating maps, providing fixes and updating AV autopilot software. They can also offer Wi-Fi hotspot services to occupants especially under the TaaS model.

### 15.2.7 Don't Copy Tesla

Every business has a role to play in the economy. Tesla's role is to be a visionary, a disrupter and a creator of non-polluting electric cars with Autopilot that consumers love to drive. Google is an innovator of a different type. Its army of computer scientists and developers expand the state of the art into uncharted areas of human endeavor to build future businesses. It is constantly rejuvenating its top cadre of innovators with bright researchers from leading universities who are working on future technologies. OEMs don't fit into either of these molds. Their management style and culture cannot match that of the technology challengers. Acquiring talent or technology-savvy companies is not a substitute for the innovation culture of the Silicon Valley challengers.

So I say to OEMs: "Be who you are". Understand your core strengths: designing the body of an AV and operating modern automated assembly lines and distribution networks. You build reliable vehicles, deliver them to consumers and service them. Negotiate with technology companies to help you with the brain of the AV. Give your AV bodies the brain they need and the user interface that allows them to speak conversationally with their owners or operators. If you deliver what you promise, you'll still control your destiny."

### 15.2.8 Don't Design Your Own Unique AV Brain

Established OEMs who have designed and built cars for decades now want to design and build their own brands of artificial brains for autonomous vehicles. Some have acquired small AI start-ups or set up R&D centers staffed by academic researchers from Stanford, MIT and other prestigious schools. This means there could soon be scores of artificial brains driving around the Silicon Valley. The industry can't sustain that.

Imagine if we had 10 to 15 different operating systems available for PCs or smartphones instead of the two or three we have today. The PC industry succeeded because there was just Windows and then Mac. The same is true of smartphones, because there were only two platforms. How could we sustain an industry with that many species of brains? I believe that the AV industry will experience a natural attrition process, and only a few robo-brains will survive. OEMs will be able to equip their AVs with an industry-standard brain framework (a sort of basic set) and then upgrade it with a unique "look and feel," giving it the OEM's specific brand characteristics.

### 15.2.9 Free OTA Updates

In the past, dealers have tried to ask consumers to pay hundreds of dollars to update their navigation maps. That model is unsustainable. Consumers will expect free OTA updates of their AV software (including updates to the core OS or the AI, or any critical security patches); upgrades of core software must not be revenue-generation tool. Look at the software industry—Microsoft doesn't charge for OS upgrades. It simply upgrades Windows 10 when we're all asleep. Tesla has already set the example with free Autopilot upgrades, only charging when there's a significant functional enhancement that was never promised at the time of purchase. The AV industry must follow that lead.

### 15.2.10 Establish Responsive Remote Support Services

AVs will have lots of problems that will require remote support, especially in the first five–10 years. AV manufacturers haven't talked about their plans in this area. In order to build consumer confidence and trust, AV vendors should announce their plans and assure customers that they'll have help if the TaaS vehicles they're in break down or get into trouble. The GM OnStar and Ford Sync networks are good starting points, but the expectations are going to be much higher when human drivers can no longer just call AAA/CAA to be towed to the next gas station; they'll be completely at the mercy of the support network.

### 15.2.11 Build Commercial-Grade, Fault-Tolerant AVs

The market expects that TaaS vehicles will operate virtually 24/7 or 16 hours a day (two shifts). These vehicles will rack up huge mileage (potentially a million miles in the five-year life expectancy of a commercial TaaS vehicle). To build a car that can stand that kind of use, we need a stringent design criterion, highly reliable and redundant components, mission-critical architecture and a fault-tolerant OS. OEMs should model their design and test philosophy similar to the aeronautical industry. Commercial aircrafts are capable of flying on auto-pilot with plenty of air space around except while landing and take-off. Yet regulators do not allow it—ultimate safety of passengers is the issue.

### 15.2.12 Support Apple CarPlay and Android Auto Enthusiastically

For several years, OEMs fought a losing battle against consumers' desire for Apple CarPlay and Android Auto in their cars for subset of functions that are considered safe and do not violate "hands free" laws in various jurisdictions. OEMs may feel that they can provide the same functionality through their own infotainment solutions, but they will likely find it difficult and expensive to build such a system, not to mention a drain on resources. A proprietary system also won't benefit from the cross-fertilization of ideas that happens in the computer industry. AV infotainment simply cannot and will not replace the personal smartphone, which is richer in functionality, has been around much longer and is almost universally used by billions of users.

### 15.2.13 Hire Savvy Staff for Dealer Showrooms

The auto industry is changing fast. AVs will bring significant changes to every aspect of the industry over the next three decades. Dealer showroom staff must also change. Staff must be AV-savvy with extensive digital sales expertise. They need to be able to communicate in a language and business style that allows them to connect with future generations of consumers.

### 15.2.14 Rejuvenate Service Organizations

Just as the outside sales staff must change, so must the service staff. Many of the problems AVs will have involve software and hardware components, which will require a different breed of service technicians. We will still need traditional services like oil changes, brake-pad replacement and tire rotation, but more often we'll need computer-whiz techno-surgeons who can handle hardware and software problems, including issues with AI. They may be hard to come by at first, but our vocational institutions and AV vendors will soon catch up and begin training this new breed.

### 15.2.15 Learn Some Tricks from the Smartphone and Telecom Industries

There's a lot that the auto establishment can emulate from the smartphone and telecom industries.

1. Telecom companies enjoy higher profit margins because much of their business support infrastructure is automated. In contrast, the auto industry uses dealer networks and human resources to provide support services. The auto industry must produce more reliable AVs using fewer components and be able to provide sales/maintenances services remotely. This will reduce reliance on visits to the dealer and will come at a lower cost—which is good for the customer *and* the OEMs.
2. Technology vendors provide powerful web-based remote support and customer service through the cloud using chat lines and bots, with real people as a last line of support. The auto industry could base its future service on that model.

## 15.3 What Should Technology Challengers Do?

Just as the established OEMs can take steps to accelerate the AV race, technology challengers can do their part to help the industry reach its destination faster.

### 15.3.1 Decide On Your Future Role—Google or Tesla Style

Technology challengers come in two flavors—Google and Tesla. Google has, so far, been focusing on building prototype AVs by retrofitting non-AVs from OEMs like Toyota and FCA with sensors and robo-autopilot software; essentially they've been concentrating on the upgrade technology required to make an AV. Just like with smartphones, Google is not interested in manufacturing cars. Margins in the manufacturing part of the auto industry are not in line with what Google is used to in rest of its business. Tesla, on the other hand, wants to be a vertically integrated, full-service auto manufacturing company. They want to control all aspects of the AVs that they build. Tesla's strategy is more difficult to execute, which shows in the company's sluggish delivery performance. It's been a long struggle and the jury is still out on whether Tesla will become a profitable venture.

### 15.3.2 Waymo: Don't Become Tesla

Should Waymo/Alphabet acquire an auto OEM and become a vertically integrated AV manufacturing company like Tesla? I don't think that's a good idea for Waymo or for the AV industry. It would mean wasting a lot of management attention and energy on a low-margin product line. Waymo doesn't have the management style or the know-how to operate an auto OEM. It would find the competition intense—Alphabet companies are used to duopoly or "triopoly" markets. Acquiring an OEM would mean Waymo would lose its ability to license its robo-autopilot software and computing expertise to other OEMs. It would appear that Waymo has heard me already because Waymo's CEO, John Krafcik, recently outlined the company's business strategy: it does not intend to build vehicles. Instead it will license its software and related integration services to auto OEMs. I think Waymo should go beyond FCA Pacifica model vehicles to run AV trials. May be FCA is more willing to accept Waymo's terms but I would like to see Waymo autopilot in more brands.

It makes business sense for Waymo/Alphabet to expand on its current course. The company could easily be a leader in building robo-autopilot that it can then license to multiple OEMs. As a systems integrator, Waymo could also help OEMs with building an AV OS framework that controls all the hardware in the vehicle through a unified OS in competition with a QNX/Nvidia/Intel OS framework. While I don't see GM, Ford Daimler licensing Waymo's robo-autopilot in the *near* future, I *can* envision a scenario where the OEMs change their strategy after realizing what an expensive and resource-intensive exercise it is to try to build their own. OEMs may also realize that Waymo has a sturdier autopilot design than the competition, with superior quality control.

### 15.3.3 Waymo: Build Modular Software—"Let OEMs Have it Their Way"

I am suggesting that instead of offering the entire robo-software as a complete package, Waymo should offer autopilot software in modular components and let the OEMs integrate them with other pieces (infotainment modules, an OS framework, etc.) that the OEMs either build themselves or license from other organizations. Each OEM can build a "best of breed" AV software framework for their unique requirements and brand it under their own name. This open, modular software-licensing concept is something Google/Alphabet companies generally use in other lines of business, and would require that Waymo not insist on an exclusive licensing arrangement with any OEM.

### 15.3.4 Mobileye and Nvidia: Stay the Course

What about Intel/Mobileye and Nvidia? Both companies have similar ambitions in the AV sector, although each of them has its unique strengths. Compared to Nvidia,

Intel/Mobileye has deeper pockets with well-recognized silicon hardware design and manufacturing facilities. I expect they will use Intel's design capability and factories to build cheaper sensor hardware. Nvidia has significant advantage in building the specialized supercomputers that are required to run AI software for the autopilot. Both companies should pursue their current strategies. It's good to have some competition in that space.

### 15.3.5 Tesla: Solve Scale and Profitability Issues

I commend Tesla for proving that an electric car with advanced AV capabilities (SAE level 3, so far) can be built. Its innovative OTA AV software updating is remarkable and needs to be emulated by OEMs as they move forward, and it has forced Detroit to become serious about electric cars. The Tesla experiment has proven that with determination, innovation and perseverance, we can bring a modernized Detroit to the Silicon Valley. However, Tesla fell short on meeting its delivery and profitability promises. The company has much to learn about the complexities of scaling assembly-line automation. It will truly be tragic for the industry if Elon Musk is unable to solve the scaling issues for the Model 3 and loses the public's trust. However difficult the problems, I believe they can be solved. Pride aside, if Tesla has to buy manufacturing expertise or acquire capacity from an innovative OEM, it will be good for the AV industry and for society's pursuit of safer and cleaner cars for future generations.

## 15.4 What Should Regulators Do?

The public and the industry are looking to regulators to guide the future of the AV world. Regulators must do their jobs and not expect the industry to self-regulate. It's their duty to represent the interests of consumers. I'm disappointed by the guidance they've provided so far. The US DOT's guidance document (see Chapter 12) does not have enough depth and detail. I suggest that the DOT and DMVs in the United States (and similar organizations in Europe and other advanced economies) should provide technological and regulatory leadership guidance as follows.

### 15.4.1 Create a Reference Design for Mission-Critical AVs

I believe that the first generation of AVs only partially accounts for the technology's mission-critical nature. In the second generation of AV design, we need to look at more advanced architectures that emphasize backup, redundancy, fail-safe and graceful degradation principles. We should use fail-safe components and software concepts, redundant sensor hardware where appropriate and non-stop computing. These concepts are used in the aircraft industry and in financial transaction processing systems. The DOT and similar bodies should create a task force with representation from the DOT, industry, academia and independent consultants to create a reference design for a mission-critical AV based on these concepts. This design should use dual components (one main and one "hot" backup) where appropriate and OS software similar to airline and financial fault-tolerant systems. Auto component suppliers like Bosch are already using dual steering components that share the load and in case of failure, provide half the capacity. Of course, cost considerations and complexity of design are factors that systems engineers need to consider. The reference design will essentially provide high-level principles, leaving it up to the industry to complete detailed designs for implementation.

If these concepts are not adhered to, we could end up with chaotic conditions on the roads when AV hardware and software fail, especially in congested cities.

### 15.4.2 Support Third-Party AV Certification Facilities

DOT-like bodies should assume overseer responsibility for third-party certification facilities like the American Center for Mobility in Michigan. Since AV certification is a regulatory function and a federal responsibility, federal bodies should initially fund such facilities. There should be well-defined penalties for AV suppliers who don't meet certification requirements and whose products are put onto the road.

### 15.4.3 Have Consumer Interests Represented in DOT-Industry Advisory Groups

While the regulatory bodies are expected to represent the interests of consumers, there should be an explicit representation of consumer advocate bodies on DOT AV advisory panels and committees.

### 15.4.4 Provide Retraining for Law Enforcement

Federal DOT and similar bodies should provide retraining and a set of tools that will assist state/provincial law-enforcement bodies to deal with the new requirements that AV disruption will introduce. There will be many situations that require new methods and knowledge.

### 15.4.5 Create a National R&D Fund

The SAE must go beyond its brief J3016 document and create an expanded version that provides detailed, explicit specifications for each and every feature of AVs—how they work and whose responsibility they are. There must be no confusion among consumers over whether a vehicle meets SAE level 3, level 4 or level 5.

Major countries such as the United States and those in Europe should create national R&D funds to support fundamental research by the academic community in areas that are crucial to the success of the industry—areas that have not yet been fully resolved. A public-private consortium can manage the fund. The cellular industry, especially in Canada, has a great example to follow—every telecom service provider has to contribute a very small percentage of its revenue to an R&D fund. It was an easy sell for providers, because the industry cannot operate without spectrum and the government owns the spectrum. I think there's a justifiable case for instituting a similar R&D fund for AVs where so much is at stake. There is a parallel here because AVs cannot operate without road infrastructure and pubic owns that infrastructure.

### 15.4.6 Amend Transportation and Liability Laws in a Timely Fashion

The industry is calling for revisions to various transportation-related laws to account for AVs operating on public roads. The whole concepts of driver, vehicle occupant, vehicle owner and mandatory driver emergency help require attention. This should be done in good time so that the industry can decide on necessary business plans. These laws will have a bearing on robo-autopilot as well.

## 15.5 What Should Cities and State/Provincial DMVs Do?

Cities and states/provinces are important AV stakeholders. They may reap huge benefits if the AV revolution happens successfully. However, they have great responsibility as well. While cities always have funding difficulties and cannot easily raise taxes for new initiatives, they should look into the following ideas to prepare their cities for AV adoption:

1. The National League of Cities in conjunction with state/provincial bodies and DOT-like bodies should look at public transportation in a holistic fashion where AVs are just one of the solutions.
2. Cities should launch initiatives to set targets for making their cities AV-ready.
3. DOT-like bodies, in conjunction with National League of Cities should develop guidelines and standards for smart cities, including things like durable lane markings and smart signs for speed limits and turns.
4. Public safety organizations should develop connectivity standards for emergency-response functions (such as fire and medical). Ohio State University is currently working on this.
5. Small municipal jurisdictions may find ride-hailing and TaaS a more cost-effective public transportation solution.
6. Platooning of transport trucks is not a good idea, as it poses a hazard that private vehicle owners and TaaS operators will hate. Imagine a platoon of six 53-foot-long AV trucks (spanning 500 feet with just small gaps between vehicles) filling up the right lane traveling at 65 mph and you are trying to pass this truck platoon or exiting the highway to deliver stuff to an important customer?

## 15.7 What Should Departments of Education and Researchers Do?

While top research universities in the United States, Germany, Italy and the United Kingdom have been doing leading R&D in AV-related areas, there's now a need to create educational programs at the undergraduate level and, more importantly, at the vocational colleges where the bulk of auto-service personnel come from. New courses will need to be developed and old courses extensively upgraded.

## 15.8 What Should Insurance Companies Do?

AVs will disrupt the auto insurance industry to a large extent. It's essential that the insurance industry start reducing premiums in recognition of the reduced risks that come with AVs. Consumers expect it. Already there is some data available showing reduced accidents with ADAS-featured vehicles. Not doing anything until large amounts of data start landing on actuaries' desks is not a good strategy, because new insurance players could start entering the business before the incumbents react.

## Summary

*I've described both strategic and tactical steps that the various industry players could take to help accelerate us into the new AV paradigm. It will require not only that innovators work together to give consumers safe and reliable AV technology that meets the highest quality standards at the most affordable price points, but also that they do so in a timeframe that all generations can adapt to. Regulators, legislators, insurance companies and cities will jump in to do their part if the first condition is met.*

---

## Citations for External References

[58] Wired magazine article on cultural shift in auto industry – https://www.wired.com/2015/11/the-auto-industry-wont-create-the-future/

# The Last Word – No, an Interim Word

I started putting together framework of the book almost 12 months back. The technology has made huge progress during the past nine years since Google started its trials in the Bay area. Both the technology challengers like Waymo and the established auto OEMs (GM, Ford and Daimler) have announced their timelines for introducing AVs on public roads. These launch dates are aggressive to say the least but PR staff will put a spin on those dates because the suppliers will have AVs on the road somewhere with some "fine print" restrictions.

The media has created a lot of awareness and expectation among the public. AV proponents have done a great job in convincing the public as well as the regulators that AVs will save lives and reduce traffic congestion. While the brave AV enthusiasts can't wait and would like to ride in fully autonomous cars tomorrow, majority of the general public is not convinced. Many recent studies and surveys are supporting the perception that AVs are not ready for the prime time or "mission-critical" task of moving passengers without a human driver. Recent fatal accidents with AVs have helped to reinforce that lack of trust.

Even the rationale for AVs is being scrutinized by serious analysts who are saying that promised benefits will not be realized during the first couple of decades and may go through a rough patch. Postponement of those benefits may leave a bad taste among consumers, infrastructure custodians and regulators.

The industry marches on, keeps on moving forward, and may have to take a breather or two at resting spots. However, this journey will continue.

Can we really write the last word on AVs? In all fairness, as our book has said repeatedly, we can't write the last word when our maiden journey has just begun. So we shall just write an interim word.

The interim word is – the industry is moving forward at a pretty good clip. We just have to be patient. We have to be careful. We have to be diligent in testing and regulations. We have to discover more. We have to make robo autopilot truly intelligent so that it can hold its own in front of its master – the human pilot.

# Index

Accidents and Deaths ......................... 64
**Amnon Shashua** ............................... viii
**Apple** .................................................. 118
Aptiv .................................................. 134
Artificial Intelligence ......................... 106
Audi .................................................... 124
Autopilot .................................... 96, 105
AV History ........................................... 40
AV OS Players ..................................... 96
baby boomers ..................................... 34
Baidu ................................................. 133
BMW ................................................. 123
**Bosch** ............................................... 134
**Cameras as Sensors** ........................ 80
Car2Go .............................................. 149
Challenges Facing AV Industry ........ 195
City Infrastructure ............................ 206
*City of the Future* ............................. 42
**CMU** ........................................... viii, 45
CMU's Navlab Project ......................... 45
**Complexity of Technology** ............. 196
Component Design ............................. 75
Consumer Perspective ..................... 156
Continental ...................................... 135
Cybersecurity Issues ........................ 205
Cybersecurity Services .................... 221
Daimler Mercedes ............................ 121
**Deep Learning** ..... vii, 71, 107, 108, 109, 112
Delphi ............................................... 134
**Dickmanns** ............... vii, viii, 46, 47, 57
Didi Chuxing .................................... 148
Domain Controllers ............................ 89
**DSRC vs. 5G** ................................... 202
Ecosystem_AV ................................... 113
Electronic Embedding ........................ 42
**Elon Musk** ................. vii, 117, 226, 230
Embedding of Roadways ................... 42
Ethics Perspective ........................... 181
**Eureka** ................... viii, 25, 46, 57, 121
Eureka PROMETHEUS .......................... 46
FCA ................................................... 130
Firefly ................................................. 51
First DARPA Grand Challenge ........... 48
Ford .................................................. 119
Francis Houdina ................................. 40
Fundamentals .................................... 31
Futurama exhibit ............................... 42
General Motors ................................ 120
Generational Gap ............................... 33
Google ....................................... 50, 117
Handoff Conundrum ......................... 198
HMI ........................................... 100, 220
Human-Machine Interface ................. 91
Insurance Issues .............................. 173
Intel ................................................. 136
J3016 ................................. 20, 22, 260
Legal Issues ..................................... 177
Levels of Autonomy ........................... 20
**LiDAR** ............................................... 79
Lyft .......................................... 147, 148
**Machine Learning** ................. 107, 108
Magna .............................................. 135
**Mapping and Localization** .............. 88
Marketing Challenges ...................... 206
**Maturity of AV Technology** ........... 195
Mercedes ......................................... 121
Mobileye ..................... viii, 25, 99, 258
Navya ............................................... 150
Neural Networks .............................. 109
NHTSA ......................................... 20, 22
NIO ................................................... 133
Nissan- ............................................ 131
Nvidia ........................... 98, 137, 258
NXP .................................................. 137
**Obstacle Avoidance** ........................ 89
Ola .................................................... 149
Operating System - AV ...................... 92
Opportunities .................................. 217
OTA .................................................. 256
**Path Planning** ................................. 89
Pedestrians and Cyclists ................. 200
**Perception** ...................................... 88
Perspective of Infrastructure Owners 170

Perspective of Innovators .................. 169
Perspective of Technology Challengers ............................................... 165
Perspectives of AV Suppliers ............ 162
platoons ............................................... 71
Polysync ............................................... 99
Preference Settings ........................... 104
**Prof. Hinton** ......................................... vii
**Prometheus** ................................... viii, ix
QNX ..................................................... 97
**Radar** .................................................. 80
Radio Control ....................................... 42
RCA Labs ............................................. 42
Redundancy Design ........................... 197
Regulators Not Ready ....................... 207
Regulatory Issues .............................. 179
rework .................................................. 33
RideCell ............................................. 153
Route Planning .................................... 87
**Sebastian Thrun** ................. vii, 26, 115
Sensor Comparison ............................. 82

Sensor Fusion ..................................... 86
Smart Highways .................................. 43
Stanford Lab Cart Project ................... 44
TaaS ............................................. 27, 141
Tesla .................................................. 176
Tesloop ............................................. 152
*Texas Instruments* ..................... 87, 136
Toyota ............................................... 128
Trivia - AV ........................................... 12
Trust ............................................ 37, 105
Uber .......................................... 146, 148
**Ultrasonic sensors** ............................ 78
UX ..................................................... 100
**V2V, V2I and V2X** ............................ 202
Volkswagen ...................................... 124
Volvo ................................................. 126
Waymo .............................................. 114
**Weather Conditions** ....................... 200
ZipCar ............................................... 150

CPSIA information can be obtained
at www.ICGtesting.com
Printed in the USA
LVHW060238210619
621933LV00003B/50/P